LIFECLOUD

THE ORIGIN OF LIFE IN THE UNIVERSE

LIFECLOUD

THE ORIGIN OF LIFE
IN THE UNIVERSE

Fred Hoyle and
Chandra Wickramasinghe

1817

HARPER & ROW, PUBLISHERS

New York, Hagerstown, San Francisco, London

FIRST U.S. EDITION

ISBN: 0-06-011954-3

LIBRARY OF CONGRESS CATALOG CARD NUMBER: 78-20167

79 80 81 82 83 84 10 9 8 7 6 5 4 3 2 1

Contents

Illustrations   7

Note on technical terms   11

Acknowledgements   12

Introduction   13

1   Our wider heritage   15
2   Myths, miracles and the origins of life   21
3   Evolution in action   29
4   Life cells   33
5   Life chemicals   41
6   The world of atoms   53
7   The stars are formed   59
8   The gas between the stars   69
9   From interstellar gas to dust   79
10 The beginnings of biochemistry   87
11 Comets: visitors from distant space   99
12 Meteorite clues   107
13 The birth of the solar system   115
14 First days on Earth   127
15 Exploring nearby planets   135
16 Planets of life   143
17 Predators and planets   151
18 Invasion from the galaxy   157
19 Communicating through space   167

Appendix 1: The chemical elements and their abundances
in the universe   177

Appendix 2: Optical isomerism of biological molecules   181

Index   187

Illustrations

Plates: Between pages 96 and 97

1 Section of carbonaceous chondrite Mokoia
2 Great spiral galaxy in Andromeda
3 Spiral galaxy in Virgo
4 Irregular galaxy: Large Magellanic Cloud
5 Elliptical galaxy in the Virgo cluster
6 Cluster of galaxies in the constellation Pavo
7 Crab nebula in Taurus
8 Milky Way in Sagittarius showing dust clouds
9 Orion nebula
10 Comet Kohoutek
11 Meteor trail
12 Close-up of Mars, across Argyre Planitia
13 Landscape of Mars showing valley with tributaries
14 Bonn Radio Telescope
15 Arecibo Observatory

Figures

1	Formation of solar system and evolution of life on Earth	31
2	Eukaryotic cell	35
3	Structures of D-glucose, cellulose and starch	44
4	'Rail-track' segment of DNA	46
5	Twisted-ribbon model of DNA double helix	46
6	DNA double helix split by action of an enzyme into two identical structures	50
7	Detailed structure of chlorophyll	51
8	Schematic models of hydrogen, helium and carbon atoms	54
9	Energy levels of hydrogen atom	57
10	Schematic views of our galaxy face-on and edge-on	62
11	H-R diagram of stars	63
12	Evolution of stars of 1 and 10 solar masses	65
13	Interstellar abundances of elements	73
14	Dimming of starlight in its passage through cosmic dust clouds	81
15	Black body spectra for various temperatures	88
16	10 micron excess in spectra of cool oxygen-rich stars	90
17	Thermal radiation from Trapezium nebula compared with calculations for cellulose	92
18	Agreement between observed infrared radiation from astronomical sources and cellulose models	93
19	Formation of pyran rings and their linkage into polysaccharide skeletons	96
20	The 2200 Ångström band due to nitrogenic ring compounds compared with astronomical data on starlight absorption	97
21	Orbit of Halley's comet	101
22	Planetary orbits in the solar system	117
23	Birth of the solar system: rotating gas sphere flattening and developing a disc as speed of rotation increases	119
24	Birth of the solar system: magnetic 'spokes' transfer angular momentum from central condensation to expanding disc	120

Figures

25 Spectrum of chlorophyll A compared with energy
distribution of sunlight at Earth's surface   129
26 Life forms in the geological record   133
27 Expanding wave-front in interstellar colonization 159
28 Mirror images of two amino acids   182
29 Mirror images of two sugars   182

Note on technical terms

Throughout this book the term 'billion' is used in the modern scientific sense of a thousand million.

Unless otherwise stated, temperatures are given in degrees Kelvin. The Kelvin scale begins from absolute zero, so that 0° Kelvin is approximately − 273° Centigrade. The Kelvin and Centigrade (or Celsius) scales always increase or decrease by the same amount and are therefore always approximately 273 degrees apart.

Wavelengths are given in Ångströms (1 Ångström = 1 ten billionth of a metre).

Acknowledgements

The authors and publisher would like to thank the U.S. National Academy of Sciences for permission to quote two extracts from an article by George Wald published in the *Proceedings of the National Academy of Sciences*, vol. 52, 595, 1964.

Introduction

Lifecloud is concerned with the astronomical basis of the origin of life. At the level of the chemical elements it has been known for some years that nuclear processes within stars are responsible for synthesizing carbon, nitrogen and oxygen from the simpler elements hydrogen and helium. More recently it has been established that within the interstellar gas clouds carbon, nitrogen and oxygen, together with hydrogen, form a wide variety of organic molecules. Astronomy therefore has a hand in the creation of life-forming material already at the organic level.

In this book we shall go beyond this and argue that astronomy is involved in the origin of life at the level of complex biomolecules, particularly the chains of sugar molecules linked through hydrogen atoms and known as polysaccharides. We shall present evidence relating not only to these chains but to another group of biomolecules, the nitrogenated ring molecules, and show how these two types of biomolecule were built from their constituent individual atoms. We maintain that the essential building-blocks of life exist on an astronomical scale in very great quantity.

We shall also argue that the interstellar cloud of gas and dust in which our solar system formed continued to add biomolecules long after the early high-temperature phase of the solar nebula and of the planetary material had been completed. Such additions of biomolecules provided the building-blocks for the emergence of still more complex forms, which eventually developed into the first living cells.

Our third contention deals with the likely site for the fitting of biomolecules into more complex forms. The Earth was a possible site, but seems less favourable than the multitude of comet-sized planetesimal bodies that must have existed over the first few hundred million years in the history of the solar system. It is also very likely that the Earth derived all its comparatively volatile materials in the atmosphere and the oceans from such planetesimal bodies. Our argument is that life arrived eventually on Earth by being showered already as living cells from comet-type bodies.

13

To set the background for these arguments, the first five chapters are concerned with various aspects of terrestrial biology. These are followed by four chapters which are largely astronomical and which lead up to Chapter 10, where evidence for the existence of biomolecules within the interstellar clouds of gas and dust is discussed. Chapters 11 to 14 deal with comets and meteorites, the birth of the solar system, and the possibility of the origin of life being cometary.

The final chapters are concerned with planets other than our own: Chapter 15 is about the other planets of the solar system, Chapter 16 is about planets moving around other stars, while the remaining ones deal with the question of the origin of life in other systems and with the possibility that eventually the human species may succeed in establishing communication with other life forms existing in other stellar systems within our galaxy.

Our wider heritage | 1

The precise moment of our emergence on this planet, the emergence of *homo sapiens*, is still a matter of dispute. On a geological timescale it must surely be very recent – no more than a few hundred thousand years ago – yet our biogenetic and chemical links extend to a very remote past, and probably to distant regions of the universe. How far into the past and how far away in space can our origins, and indeed the origins of all life, be traced? How did life reach our own planet – and possibly others? These are the main questions that we shall discuss in this book.

We are inescapably the result of a long heritage of learning, adaptation, mutation and evolution, the product of a history which predates our birth as a biological species and stretches back over many thousand millennia. Yet our day-to-day experience of heritage and history is relatively 'local' in both time and space, since to us as individual members of a species it is our immediate filial and cultural links with other individuals or groups that are the most self-evident. It is a relatively simple matter to trace connections with other contemporary groups who share a common background of learning, or with individuals or groups in our immediate past. We could, each of us, trace salient features of our pedigrees, linking ourselves with far-flung parts of the globe through past migratory movements, or we might link ourselves with individuals associated with certain historical land-marks – for instance the Napoleonic wars. Biologically this type of local pedigree is not especially interesting, but a more radical process of ancestor-tracing can lead to a more basic and exciting discovery about not only our own ultimate origins but the origins of all life.

Going further back, we share a common ancestry with our fellow primates; and going still further back, we share a common ancestry with all other living creatures and plants down to the simplest mi-crobe. The further back we go, the greater the difference from exter-nal appearances and behaviour patterns which we observe today. Always and everywhere, however, a biochemical unity has persisted, and basic similarities of physiology have spanned very wide ranges of species. In our own physiological responses to imminent danger, for example, we share with many different species the same biochemical responses to fear. An increased secretion of adrenalin has the effect of preparing us for the fray by instantly increasing blood pressure, the heart beats faster, and the glucose level in our blood rises. This is just one of the many tricks that were learnt for self-preservation quite early in our biological evolution.

When it was published in 1859 Darwin's *On the Origin of Species* met emotional opposition from almost every quarter. It was anathema to all those who held the firm belief that separate species of plants and animals were fixed and immutable. There were also deep religious and philosophical objections to accepting an idea which equated man with the very lowest forms of life on this planet. Darwin proposed that higher life forms must necessarily evolve from the lower. In this book we shall take the basic concept of Darwinism still further back and come to the conclusion that a pre-Darwinian molecular chemical evolution preceded the emergence of the first living species.

Darwin's theory, which is now accepted without dissent, is the cor-nerstone of modern biology. Our own links with the simplest forms of microbial life are well-nigh proven. From a biochemical point of view the difference between man and microbe is comparatively tri-vial. At a most rudimentary chemical level, life in all its varied shapes and forms simply involves the interaction between two groups of biochemical substances, the nucleic acids and proteins. The nucleic acids are themselves constructed from just one sugar, four bases – adenine, guanine, thymine and cytosine – and a phosphate, while proteins contain about twenty-one separate amino acids. The myriads of possible arrangements and re-arrangements of these twenty-seven or so basic substructures make up the wide diversity of life forms on our planet.

A question of much deeper fundamental significance relates to the origin of these basic biochemical building-blocks of life. Once these fewer than thirty biochemical building-blocks have formed and become assembled to make up a primitive life system, Darwinian evolution takes over and everything else follows. Organisms of ever-increasing complexity and sophistication develop by a process of natural selection, which goes on slowly, almost imperceptibly, but nevertheless relentlessly through epochs of changing physical conditions on our planet.

But what of our ultimate chemical and biochemical genesis? And how is this genesis connected with the universe at large? Answers to these questions will form the central theme of our book. There is a fairly obvious sense in which the ultimate genesis of all life-forming matter is intimately associated with astrophysical processes. This is at the level of the architecture of atoms. All the individual atoms of carbon, nitrogen, oxygen and metals which make the molecules of living systems were processed from the simplest atom, hydrogen, by nuclear reactions which took place in the interiors of stars or in the material expelled from stars of large mass when they exploded as supernovae. So much at least is incontrovertible; there is no doubt that organization of matter to at least this level of complexity must have occurred on a cosmic scale.

Up until a few years ago, the next stage of organization, which was a molecular organization involving associations of single atoms to form the complex building-blocks of life – the components of nucleic acids and proteins – was supposed to be an entirely terrestrial affair. Precisely how this organization took place was never very well defined, but the most widely held belief was that these molecular building-blocks were re-arrangements of simpler combinations of atoms, such as are present in water, carbon dioxide and ammonia. The energy required to accomplish these re-arrangements was thought to be supplied by the action of electric discharges in lightning or through ultraviolet light received from the Sun.

New clues which have a profound bearing on the question of the origin of life have only very recently become available. The new

17

information has been acquired by an interplay between several disciplines: astronomy, laboratory studies of meteorites, physics, chemistry and of course biochemistry. It is now clear that a long heritage of pre-Darwinian molecular evolution occurred in a cosmic rather than a purely terrestrial context, predating by many billions of years the formation of the first organisms on Earth.

Emissaries of our biochemical heritage still arrive on our planet in the form of meteors and meteorites. Some hundred or so metric tons of matter plunge daily into the Earth's atmosphere. Much of this material is in the form of fine particles, meteors, many of which burn up before they ever reach the ground, but there are also larger objects, meteorites, which apart from a scouring off of their outer layers arrive on the ground more or less intact. As we shall see later, a certain class of meteorite known as the carbonaceous chondrites (Plate 1) is rich in organic chemicals.

Several times a day the Earth is hit by a body large enough to survive its journey through the atmosphere. Typically, such a meteorite body is about as big as a football. As it travels through the upper regions of the atmosphere the spectacular brilliance which we see is due to a continual sparking off of surface layers by high-speed collisions with gas atoms. The brightness of a meteorite fades away rapidly while it is still some ten miles above the Earth's surface, indicating that the speed has already been significantly reduced by gas friction. The final landing is then relatively 'soft', and meteorites often bury themselves only slightly below the Earth's surface. Because much of the Earth's surface is ocean, and because much is sparsely inhabited, meteorites are not easily recovered. Every year about ten are seen to fall and are tracked down and picked up. The total number of meteorites collected in this way is no more than about 1500 and those are carefully kept in museums and laboratories.

There have been many attempts to determine ages of meteorites by methods of radioactive dating. At least some types are very primitive and were pieced together almost certainly before the formation of the Earth's crust and perhaps even before the birth of the Sun as a full-fledged star.

There are also much larger meteorites such as one that gave rise to the meteorite crater in Arizona. This crater is nearly six hundred feet deep and about four thousand feet in diameter, and it was probably caused some tens of thousand years ago by the impact of a meteorite weighing well over a hundred tons. More recent arrivals of similar large meteorites have occurred: in 1908 and 1947 large meteorites possibly of cometary origin struck ground in central and eastern Siberia.

Where do the fine particles which enter our atmosphere come from? Most of them come at well-defined times of the year in the form of meteor showers which almost certainly originated in comets. Comets are probably derived from a reservoir of bodies which surround the whole of our planetary system out to a distance of one tenth of a light year. Kilometre-sized cometary nuclei with an icy composition are thought to plunge from time to time into the inner regions of the solar system. An example of such a comet was Comet Kohoutek, which passed through the inner solar system in the year 1973. Its present orbit is estimated to carry it round the Sun every 75,000 years. As it passes close to the Sun, heat causes volatile gases and solid particles to spray out from the cometary nucleus, and it is this spray of matter directed radially outwards from the Sun by solar pressure which makes the spectacular display of a comet. Although Kohoutek's visual display was a disappointment to most observers, the discovery of organic molecules in the cometary 'spray' will have a bearing on our story. In comets such as Kohoutek we are witnessing an exchange of material between the outer and inner parts of the solar system, an interchange which may well serve to connect us with still more distant parts of the galaxy.

Comets with long periods of revolution about the Sun have their orbits gradually perturbed by the gravitational action of the more massive planets, particularly Jupiter. These perturbations can have the effect of bringing comets into orbits of progressively shorter period, until they eventually become incorporated in the belt of asteroids – small planets revolving around the Sun – which lies between the planets Mars and Jupiter, and from which meteorites are thought to originate. As comets cross the inner regions of the solar system they leave in their wake streams of fine particles which account for the seasonal meteor showers. When the Earth crosses these streams at well-defined times of the year it picks up a share of cometary debris at each crossing.

In earlier geological epochs – in the youth of our planet about four and a half billion years ago – the rate of capture of cometary and meteoritic matter must have been much greater than at the present time. The reason for this is that the outermost planets in the solar system were then still being formed from billions of cometary-type objects, of which some would have crossed the Earth's orbit, while others would have plunged to the Earth's surface. We consider it almost certain that the Earth acquired all of its volatiles including water, carbon dioxide, ammonia, rare gases and complex biochemicals by this process. The origins of biology, involving the assembly of complex biochemical building-blocks of life, occurred not on Earth but in far-flung parts of the galaxy. Terrestrial biology was mainly concerned with the assembly of prefabricated biochemical structures, rather like piecing together the several parts of a child's construction kit. It is now virtually certain, moreover, that similar experiments in biological assembly occurred on innumerable occasions in many other places in the universe.

Myths, miracles & the origins of life | 2

Anyone who puts forward a revolutionary new theory about the origin of life has to take account of a formidable array of rival theories, some of which are scientifically more plausible than others. Currently the most widely accepted view about how life first started on Earth is the 'primeval soup' theory. How was this theory reached, what are its antecedents and how tenable is it in the light of modern discoveries in astronomy and biology?

The book of Genesis epitomizes a basic view shared by many religions and philosophies in diverse cultures throughout the world:

> And God said, Let the waters bring forth abundantly the moving creature that hath life, and fowl that may fly above the earth in the open firmament of heaven. And God created great whales, and every living creature that moveth . . . (Genesis 1 : 20–21).

Although men have pondered on the question of the origin of life for many millennia, all metaphysical solutions which have been proposed converge in one regard. They embody some arbitrary fiat of creation, one that implies an inexplicable miracle of one type or another. This miracle may or may not involve the intervention of a named supernatural being – such as God in the Old Testament – but, by some form of definition, the event of creation is deliberately placed beyond the descriptive scope of empirical science.

There is undoubtedly a fairly wide spectrum of belief on this matter, ranging from creation *ex nihilo* to transmutation of non-living to living material, but when it comes to fundamentals the differences of attitude even between widely separated cultures of the world are surprisingly small. In Vedic as well as Buddhist scriptures, beliefs concerning the nature of life are moulded by the doctrines of karma (destiny) and rebirth. Buddhism does not postulate a divine creator: the assertion is that the universe has no absolute beginning or end and that the creation and destruction follow one another in recurrent cycles. Belief in rebirth also implies a unity of all forms of life, human as well as non-human, and admits the possibility of transmutation from one form to another.

Many ancient civilizations, including the Babylonian, believed that the first life forms came from mud and slime in riverbeds, rather in the way that life is now considered to have emerged from the primeval soup. Even before microscopes were invented, the worlds of the 'living' and 'non-living' must have appeared grossly different to any perceptive observer and would have seemed to be separated by an almost unfathomable abyss. The extremely wide range of living things, large and small, plants and animals, is pervaded by an intangible elusive 'essence of life'. The world of the living is characterized by profusion, growth, mobility, change, responsiveness and a vaguely definable quality which could be called 'consciousness'.

Philosophers throughout the ages who considered the problem of the nature of life were broadly divided into two groups: idealists and materialists. The former group invariably postulated a supernatural entity which was inaccessible to empirical science – the 'psyche' of Plato or the 'divine spark' of various religious leaders. The second group of philosophers, very much in the minority, maintained that life was ultimately reducible to matter, a view which was closer to an empirical scientific explanation. In all the major cultures of the ancient world, however, such as China, India, Babylon and Egypt, the main beliefs about the origins of life were inextricably woven into religious legends and myths. Transformations of non-living to living matter and vice versa were almost invariably the unquestioned prerogative of gods and demons. Attempts at more rational explanations were few and far between.

Of all the scientific or quasi-scientific theories of the origin of life, the theory of spontaneous generation dominated human thought for the longest time. With relatively minor variations this theory asserts that living systems emerge from non-living matter under certain conditions spontaneously, without the action of a supernatural creative power. It is easy to see how this view came to be so widely held and so strongly entrenched. The sight of maggots emerging in great profusion from decaying meat, and of worms, flies and other creatures coming from slime seems to constitute a *prima facie* case in support of this theory. The Greek materialist philosopher Democritus (460–370 BC) maintained that the world was composed of an infinite number of uncaused and eternal atoms moving randomly in a void. The spontaneous generation of life from water and slime was held to occur as a result of an accidental meeting of 'atoms' of moist earth with 'atoms' of fire. Aristotle (384–322 BC) and his followers proposed that fireflies emerged from morning dew and that many types of small animal arose from the mud at the bottom of streams and ponds by a process of spontaneous generation. This belief was held almost without dissent by scientists and philosophers right through to the end of the seventeenth century AD. In *Antony and Cleopatra* Lepidus tells Mark Antony:

> Your serpent of Egypt is born of the mud, by the action of the Sun, and so is your crocodile.

Adherents to this belief included Newton, Harvey and Descartes.

Francesco Redi, an Italian physician and poet, was among the first to present a serious challenge to the theory of spontaneous generation. In 1668 he studied the development of maggots and flies on rotting meat and showed that maggots do not form if flies are kept away. After an extended period of claims and counterclaims the theory of spontaneous generation was finally discarded in the second half of the nineteenth century. This was mainly due to the work of Louis Pasteur. In 1860 the French Academy of Sciences offered a prize for a contribution that might throw new light on the issue of spontaneous generation. Pasteur responded to this challenge with an investigation which was to prove one of the cornerstones of modern biology. He was able to show convincingly that microorganisms which grew in certain media came from microorganisms in the air, and not from the air itself.

In 1876 investigations by an English physicist, John Tyndall, provided further support for Pasteur's findings. Tyndall devised an experiment which involved the scattering of light to show that air has the ability to carry particulate matter (including microbes), and that when clean air is in contact with media capable of supporting microorganisms no such organisms will grow. The final demise of 'spontaneous generation' came when Pasteur showed that parent microorganisms can regenerate only their own kind. This was convincing proof that microorganisms are indeed living creatures in their own right.

Pasteur's discoveries about the nature of microbes, together with Darwin's theory of the evolution of species, provided the basis for modern biology. A hitherto unrelated body of factual knowledge which had been meticulously collected by earlier generations of naturalists began to fall instantly into a logical pattern. The following comment from Darwin's *Notebooks on Transmutation of the Species* is of interest:

> On the average every species must have some number killed year with year by hawk, by cold, &c. – even one species of hawk decreasing in number must affect instantaneously all the rest. The final cause of all this wedging must be to sort out proper structure. . . . One may say there is a force like a hundred thousand wedges trying to force every kind of adapted structure into the gaps in the oeconomy of nature, or rather forming gaps by thrusting out weaker ones.

Evolution, mutation and adaptation provide causal links from the simplest microbe to the highest form of animal life, but what could be said of the very first microbe? On the basis of recent knowledge concerning the ubiquitous nature of the genetic code, it may be possible to assert that the origin of life, if it took place on the Earth, was a unique and singular event. Yet this original event is concealed in something of a tinderbox of mystery. A 'mystical' spontaneous generation has been implicitly conceded for the initial formation of a biological system from inorganic matter.

The question of the origin of life from inanimate matter was taken up again by physicists, chemists and biologists in the first few decades of the present century. The need for an empirical approach within the scope of modern science is well recognized, though a large part of the myth and mystery which pervaded religious and philosophical attitudes of earlier epochs is present even in the contemporary scientific answers which have been proposed.

The English biologist J. B. S. Haldane and the Russian biochemist A. I. Oparin were independently the first to recognize that life must logically have had an inorganic chemical ancestor. They suggested that the first life arose from inorganic matter under conditions which were thought to have existed on a primitive Earth. The initial premise is that there could be no primordial interstellar biochemicals, a premise which we shall later show to be wrong. In the Haldane-Oparin scheme the starting chemical system is a mixture of simple inorganic gases (molecular hydrogen, methane, ammonia, water) in atmospheric clouds or dissolved in oceans. Such molecules colliding with one another cannot form complex organic molecules. Ultraviolet light from the Sun was therefore invoked, directly or indirectly, to break up the inorganic molecules into 'energized' fragments. These 'energized' fragments were then supposed to react to form prebiotic molecules, such as might be required for the start of life.

The principle of this process is not in question, but there are doubts about the primitive Earth conditions which were needed to make the scheme feasible. Under present-day Earth conditions most of the ultraviolet light from the Sun is shielded by the ozone in the upper atmosphere, so there is little 'energizing' radiation now reaching the surface. Moreover, the chemical composition of the atmosphere at present is such that it oxidizes, as is evident from its ability to rust iron. This oxidizing property is unfavourable for the persistence of complex organic molecules, and no significant yields of biochemicals could form and be retained even if there were a great increase in the incidence of energizing events such as thunderstorms. To avoid this difficulty it has been argued that the primitive Earth atmosphere was 'reducing' rather than oxidizing; in other words the atmosphere had a tendency to remove oxygen rather than add it. The composition which is usually suggested is one that is dominated by molecular hydrogen, with smaller quantities of water, methane and ammonia. We shall argue in Chapter 14 that this initial composition is unlikely, for the early Earth atmosphere was largely made up of carbon dioxide and water, and although this mixture was different from today, it was nevertheless an oxidizing environment and not a reducing one. The primeval soup cannot form in this case.

Several laboratory experiments were carried out with a view to testing the Haldane-Oparin hypothesis. They are all technically impressive, but we doubt their relevance to primitive Earth conditions or to the start of life. In 1953 the American chemist Stanley Miller showed that high yields (up to three per cent) of amino acids could be formed by sparking a mixture of molecular hydrogen (H_2), methane (CH_4), water (H_2O) and ammonia (NH_3) continuously for one week. More recently the Ceylonese chemist C. Ponnamperuma and his colleagues ingeniously extracted traces (about 0.1 per cent) of some of the components of nucleic acids after irradiating similar mixtures with an electron beam. Some people claim they have produced trace quantities of sugars in other experiments of this type. These are undoubtedly triumphs for experimental biochemistry, but they do not constitute proof of the Haldane-Oparin soup theory as is usually claimed. The proof of the soup is in the eating!

Apart from the highly questionable prerequisite of a strongly hydrogen-dominated primeval atmosphere, there are other unanswered questions. Could enough energizing radiation penetrate to give large enough yields of all the biochemicals? Why are only very few molecular units (monomers) used in biology, whereas many more chemicals form in the soup-simulation experiments? Could all the basic biochemicals form with the same starting substances and under the same set of conditions? If the Earth is being showered with precisely the right biochemicals for life, as we know it to be, why choose the more complicated explanation?

It is doubtful that anything like the conditions which were simulated in the laboratory existed at all on a primitive Earth, or occurred for long enough times and over sufficiently extended regions of the Earth's surface to produce large enough local concentrations of the biochemicals required for the start of life. In accepting the 'primeval soup theory' of the origin of life scientists have replaced the religious mysteries which shrouded this question with equally mysterious scientific dogmas. The implied scientific dogmas are just as inaccessible to the empirical approach. One hundred years ago Pasteur declared to the French Academy that 'the theory of spontaneous generation will never recover from this mortal blow'. We can see it revived now only in a somewhat different context – spontaneous generation, not of fireflies from dewdrops, but of prebiotic molecules and primitive life in thunderstorms!

It will be one of the main purposes of this book to offer a totally different point of view concerning the origin of prebiotic molecules and of life, a point of view that will turn out to integrate life into a wider astronomical background. Before coming to things that are new, however, it will be as well to extend our present discussion to the much less controversial problem of how life evolved once it had started. This we shall do in the next three chapters.

Evolution in action | 3

With his theory of evolution, Darwin was able to unify biology by making a coherent system out of an apparent jumble of facts. The numerous life forms on our planet are readily separable into species, and there is of course a multitude of such species. Within a given species, organisms resemble one another more than they resemble members of other species, and almost without exception organisms breed only amongst their own species. Why do species differ one from another? Is there a connection between present-day organisms and fossils found in the Earth's crust? How do species alter in response to climatic as well as ecological changes in their environment? Evolution provided answers to these questions.

During his epic voyage on H.M.S. *Beagle* Darwin's belief in evolution arose mainly from four sets of observations:

(1) The presence of related but different species in adjoining areas of continents.

(2) The similarity of structure of modern and fossil life forms in the same areas.

(3) The resemblance of species on isolated islands to those on the nearest continents.

(4) Differences between species on closely adjacent islands in the Galapagos archipelago in relation to their modes of life and feeding.

All these facts can be readily explained if species were not individually created but are descended with continuing modification from common ancestral forms. Mutations occur in a random way, but natural selection involves the tendency to perpetuate those mutations which confer a better survival or reproductive value on their possessors. The ecological conditions of the environment at any stage provide discriminants to select those mutations which are advantageous to survival from those which are not. All living organisms are thus adapted to their several modes of life, and each is adapted to the peculiar characteristics of its own ecological niche. Darwin himself did not claim to provide proof of evolution. What he did claim was that, if evolution occurred, many empirical facts which are otherwise difficult to comprehend would be explained.

There is evidence which shows that the evolution of life on Earth has continued for at least three billion years, and that it still continues. The ancient fossil record is to be found in sedimentary rocks – in other words, rocks which are formed at the bottom of the sea and of lakes. Sedimentary deposits are formed continuously by weathering of rocks which are exposed to air and moisture, a process which results in fragmentation of surface materials. Wind and water then transport the fragments to lower levels in lake bottoms and sea beds. Sedimentary deposits also include by-products which come from biological activity and consist of biochemicals generated by living organisms, and of microscopic and macroscopic fossils. All this is laid down and compressed to form stratified layers of rock over long periods of geological time.

Sedimentary rocks form a relatively thin layer over the surface of the Earth and have an average depth of about two and a half kilometres. This layer has a gradient of composition and carries important information on our past history. Not long after the solidification of the Earth's crust about four and a half billion years ago, the weathering processes involving running water and air currents were at work, and the sedimentation process began. Dating of individual strata is possible with the use of a variety of techniques depending on certain types of radioactivity. The oldest sedimentary deposits are found near the Swaziland border in central Transvaal. The Barberton Mountain Land here includes a few hundred square miles of the most precious geological territory, including the Onverwacht rocks which were deposited about 3.7 billion years ago.

Fig. 1 shows a timescale for cosmic events and the evolution of life on Earth. The very earliest evidence of living organisms is found about 3.1 billion years ago in the Fig Tree Chert, which is a dark flint-like rock in the Fig Tree system, also of the Barberton Mountain Land. Such evidence is seen in electron micrographs as structural forms which are similar to modern bacterial colonies. If these structures are indeed relics of ancient bacteria, then it is reasonable to conclude that organisms with quite complex metabolic systems existed as far back as 3.1 billion years ago. These structural identifications, which were pioneered by C. D. Walcott in 1883, have recently received confirmation by more subtle chemical studies. If the structures seen in electron micrographs are truly biological fossils, it is important to look for trace quantities of certain biochemicals associated with these structures. In many instances these chemicals have now been found.

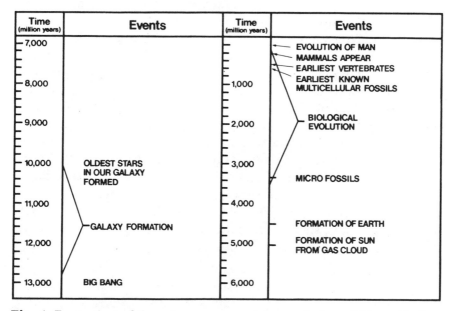

Time (million years)	Events	Time (million years)	Events
7,000			EVOLUTION OF MAN MAMMALS APPEAR EARLIEST VERTEBRATES EARLIEST KNOWN MULTICELLULAR FOSSILS
8,000		1,000	
9,000		2,000	BIOLOGICAL EVOLUTION
10,000	OLDEST STARS IN OUR GALAXY FORMED	3,000	MICRO FOSSILS
11,000		4,000	
	GALAXY FORMATION		FORMATION OF EARTH
12,000		5,000	FORMATION OF SUN FROM GAS CLOUD
13,000	BIG BANG	6,000	

Fig. 1 Formation of the solar system and the evolution of life on Earth

The more recent fossil record, up to about six hundred million years ago, is better documented. This is the period when living creatures with refractory structures – hard shells and bony skeletons – appear. The interpretation of evolutionary changes becomes more straightforward. But the claim that Darwinian evolution alone could account for the vast elaboration of species as evidenced in the paleontological record is not well founded. The great surges of evolution such as occurred after the geological cretaceous period present a great difficulty for the usual Darwinian theory. It would appear necessary that natural selection operated against a steady increase of genetic information at an average rate of about one gene per thousand years. We ourselves have argued that this gene elaboration process results from the effect of continuing viral invasions from outside.

The earliest primates appeared on the Earth at some rather ill-defined moment about sixty-five million years ago. The predominance of our immediate ancestors among life forms on this planet has therefore been restricted to a very small proportion of the total time for which the Earth's crust has existed. The earliest tangible evidence for a terrestrial life form dates back 3.1 billion years to the Fig Tree Chert. This, however, almost certainly cannot represent the ultimate origin. The evolution of primitive life forms out of some sort of pre-biotic material must have taken place between 4.5 and 3.1 billion years ago.

So much for evolution in action reckoned on geological timescales. But biological evolution is of course an inexorably continuing process, and even during the twenty years of Darwin's original observations, evidence for evolution in certain species became manifest. In very recent experiments, scientists have been able to track down mutations in bacterial strains over much shorter timescales than this. While, therefore, Darwin's insights have been of immense value in our search for the origin of life, it is necessary to look at aspects of life, matter and evolution with which he did not directly deal. In the following chapters we shall observe the world of cells, the chemicals of life, and the atomic make-up of matter.

Life cells | 4

The stage we have now reached in evolution enables us to explore the universe, our own planet and the living cell which is the microcosm within us. The discovery of nebulae by Sir William Herschel (1738–1822) and the discovery of the nucleus of living cells by Robert Brown (1773–1858) were almost contemporaneous. The two discoveries were made possible by advances in related experimental techniques – the construction of reflecting telescopes in one case and microscopes in the other – and they heralded new epochs in astronomy on the one hand and biology on the other.

In the eighteenth and nineteenth centuries a synthesis of discoveries in such widely separated fields as astronomy and biology could scarcely have been imagined. Yet the argument we shall develop in this book will show that an understanding of the origins of life can result only from a combination of these two endeavours. Herschel's dark nebulae turn out to be places where the construction of biological molecules began, and Brown's cells the places in which these molecules were further elaborated to develop into the remarkable range of properties which we call life. Before looking at the basic chemicals of life and its mechanics at a molecular level, we shall see how there is an elaborate architecture of living systems at a microscopic cellular level.

Only a few types of living systems existing here on Earth consist of single cells – unicellular organisms. The vast majority of living organisms – plants and animals – may be seen as very well-organized collections of cells. Cells are to a greater or lesser degree self-reliant, acting as separate living entities in their own right. They also differ in their size, composition and structure, as well as in the functions for which they are individually adapted. The range of size spans dimensions of less than a thousandth of a millimetre to several hundredths of a millimetre. A typical human being contains some million billion separate cells. The cells in a single such assembly are in communication, either directly or indirectly, with the external physical environment. In organisms as complex as ourselves there are many types of cell, each type performing a highly specialized and well-differentiated individual function. Thus brain cells, liver cells and red blood cells are all adapted in slightly different ways to perform somewhat different roles. Every one of them, however, contributes to the overall biological function of the organism.

The ability of a cell to replicate itself is a necessary prerequisite for life. Furthermore, multicellular organisms must be able to reproduce the large-scale organizational structure as a whole if the system is to be fairly described as 'living'. Replication and reproduction of a cell or of a biological organism in general involves assimilation of nutrients in the form of simple molecules from the outside environment. The organism's whole genetic and enzymic machinery then converts these simple molecules into more complex molecules, which are further built into structures adapted to the functioning of the particular organism. By way of contrast, a molecular system which can reproduce itself only by the assimilation of highly complex substances is not usually considered living. This is the case for viruses.

In terrestrial biology there are two fundamentally different types of cell which are associated with living systems. These are the prokaryotes and the eukaryotes. Eukaryotic cells have a well-defined centre or nuclear region surrounded by a nuclear membrane, while the prokaryotes do not. Eukaryotes also have much more complex and sophisticated substructures (miniature organs) which are not present in prokaryotic cells. The prokaryotes are confined to the blue-green algae, bacteria and to a group of mould-like systems known as the actinomycetes, while eukaryotes encompass all other cells. Fig. 2

shows a representation of an electron micrograph of a eukaryotic cell. This is seen to have a large degree of internal structure and organization, with specialized functional units tucked away in discrete compartments. The structural differences between these two types of cell probably reflect an evolutionary sequence from one to another, the simpler prokaryotic structure evolving into a eukaryotic one. It is interesting that the fossil forms of the earliest bacteria and algae discussed in the previous chapter can almost certainly be explained in terms of a prokaryotic structure.

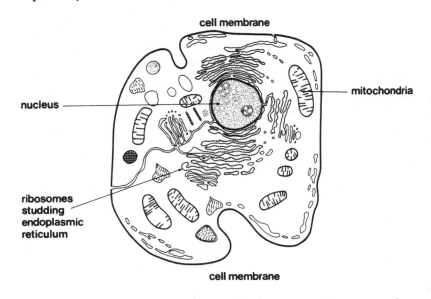

Fig. 2 Representation of a eukaryotic cell, showing major structural components

The eukaryotic cell has an elaborate structure made up of the nucleus, the cytoplasm (which occupies the main volume of the cell) and the membranes which surround the nucleus as well as the cell as a whole. The overall chemical composition of cell material is typically about seventy per cent water, twenty per cent protein, four per cent nucleic acids, with the remainder occurring as lipids (fatty acids), carbohydrates and other trace constituents. Intricate subdivisions exist within both the nucleus and the cytoplasm, and these substructures play the role of 'mini-organs', each having well-defined functions in the cell's metabolic processes. The nucleus itself is divided into nucleoli and chromatin, which together provide the genetic equipment required for the regeneration of the entire cell.

The cytoplasm, which is relatively large and complex, contains quite a number of subdivisions. The centrosomes play an important part in cell division, while the mitochondria, of which there are several hundred in a typical cell, have a crucial role in intracellular energy exchanges including the processes of photosynthesis in plants and respiration. Mitochondria are in many respects self-contained 'miniature cells' resembling the prokaryotes. Many biologists believe that they were bacteria which became incorporated into eukaryotic cells early in evolution. They are self-reproducible and have their own nucleic acids as well as enzyme-synthesizing systems. The ribosomes, which are small granules comprised of RNA-protein complexes, participate in the making of proteins. The endoplasmic reticulum is a communication system of canals which carries material within the cell. The surfaces of some of these canals are studded with ribosomes and consequently support protein synthesis. Other important parts of the cytoplasm are the lysosomes, which are membrane-bound packets of digestive enzymes, and the Golgi apparatus, which is involved in the formation of lysosome and secretions. The entire cell is surrounded by a membrane which is some ten to twenty atoms thick, made up of proteins and lipids (fats) in ordered layers. The cell membrane is a porous structure which allows matter to flow back and forth into and out of the cell.

Living cells on our own planet can therefore be said to serve several purposes. They store, transform and use energy, and they synthesize complex molecules such as proteins by interacting with a surrounding medium which is largely inorganic. They divide and replicate according to a well-defined genetic prescription, and they can adapt in response to changes in the external environment. Among all these processes in cell biology the most fundamental process is of course the primary input of energy, and it is this process which governs all other processes involving exchanges of chemical and electrical energy.

How do terrestrial organisms absorb energy from the outside world? Some of them – known as autotropes and including purple bacteria, blue-green algae and all green plants – absorb energy directly from sunlight. The cells of autotropic organisms incorporate mini-organs known as chloroplasts, which contain chlorophyll, the green substance of plants. The chlorophyll molecule has the property of absorbing sunlight and storing this form of energy in its chemical bonds. The stored energy can then be used to form molecules of glucose from molecules of water and carbon dioxide:

Water (H_2O) + Carbon dioxide (CO_2) + Chlorophyll + Sunlight $\rightarrow$ Glucose ($C_6H_{12}O_6$) + Oxygen (O_2) + Chlorophyll

In this reaction the chlorophyll acts as a catalyst by providing a storage battery for solar energy and releasing the energy to bring about a chemical reaction that is vital to all life. As we shall see, the origin of the chlorophyll molecule could have an important bearing on the question of the origin of life itself. Glucose (starch is a long string of glucose molecules) serves as a source of stored-energy which can be readily assimilated by complex organisms. This stored energy can be released in reactions with oxygen, leading to the formation of carbon dioxide and water, as well as in fermentation reactions, with the liberation of carbon dioxide and the formation of alcohol. Autotropes are therefore entirely self-sufficient. All the nutritional substances necessary for their function and regeneration are made from purely inorganic material by the direct use of solar energy.

Other terrestrial organisms – known as heterotropes and including the vast majority of terrestrial bacterial species and all animals – are unable to synthesize all their necessary nutritional substances from inorganic materials. Heterotropes directly or indirectly parasitize on autotropes, mainly in the form of plants. Heterotropes perform all their useful work by releasing energy held in complicated molecules such as starch or glucose which are contained in plants.

Starch and glucose from plants ($C_6H_{12}O_6$) + Oxygen from air ($6\ O_2$) $\rightarrow$ Carbon dioxide ($6\ CO_2$) + Water ($6\ H_2O$) + Energy

This process involves respiration, a function which in individual cells is taken care of by mitochondria.

We know that any individual cell of a living organism is itself alive in its own right. In a complex organism, of course, there are many different types of cell performing separate functions. Some simple life forms, however, are single-celled (unicellular) and in the well-known example of the amoeba, as indeed in any other cell, a couple of thousand or so chemical reactions take place. Whatever the cell, whether it is part of a single-celled organism or of a multi-celled animal as complex as a primate, the totality of these reactions results in the cell's harmonious life. In every cell the chemical system is well ordered, and not in chaos, only because of the presence of a very special class of compound known as enzymes. These are proteins which control accurately the speeds of the chemical reactions. Specific enzymes are concerned with different cell functions, and this means that the same types of enzyme carry out the same roles in very diverse organisms. This is a striking example of an evolutionary link between different levels of living things.

The evolution from amoeba to man may be understood as a progression towards increasingly refined internal communication within organisms combined with a steady increase in the total content of genetic information. There are obvious survival advantages for cells to gang together into well-knit organizations, and organizations which develop better inter-cell communication systems will be favoured – they could be better predators. Nerve cells are cells which are specially adapted to respond quickly to stimuli. Sponges, which consist of a colony of cells, are probably the simplest of all multicellular creatures. The separate cells in a sponge coordinate in only a limited way, for example to force water and nutrients through the body. The colony is therefore better able than the single-celled amoeba to exploit its environment, but its coordination is somewhat limited because it does not have a well-defined nervous system. A jellyfish has a diffuse nervous system, while the common hydra has a mesh of nerve cells which provides good sensory coordination over the whole body. In larger animals including ourselves an exceedingly intricate nervous system coordinates stimuli from all parts of our bodies through the brain, which responds in a way that determines our behaviour patterns. Nerve cells and the brain are therefore the key to our higher intelligence.

We have so far described life forms which are ultimately contained within cell membranes. Non-cellular life forms – fungi and viruses – also exist on the Earth at present. Although there is no evidence for such a life form in the terrestrial fossil record, there is every reason to believe that non-cellular life forms preceded the evolution of prokaryotes. From the standpoint of molecular evolution one could argue that the basic characteristics of life were first embodied in fairly simple macromolecular structures. The advantages of survival, conferred on those life forms that used cell membranes as distinct from those that did not, may have led to the rapid extinction of the more primitive structures.

Viruses consisting of one or more strands of nucleic acids (together with proteins) occupy a no-man's-land between the 'living' and 'non-living' worlds. Although in their organization and structure viruses are much less complex than cells, they perform functions which are in fact highly sophisticated. For their survival, however, they depend on the pre-existence of cellular organisms, and the reproduction of viruses occurs through complex interactions with cells.

Life, then, is based on the very fundamental but all-pervasive architecture of the cell. Cells are essential to life as we know it, from the simplest to the most complex organisms. Even if sophisticated life forms might conceivably exist elsewhere without any type of cell structure – and there is no evidence for their existence – the physical and chemical make-up of cells will have an important bearing on the argument for the extraterrestrial origin of living things within our direct or indirect experience.

Life chemicals | 5

New discoveries in biochemistry have given us information about the basic chemicals in living things and enable us to go some way towards identifying the very essence of life. If we are to look for the ultimate source of living things, it is necessary to see what is known about the constituents of life before asking whether it has extraterrestrial origins.

The living cell contains many different chemical substances distributed among its several components. The single most abundant substance which makes up about seventy per cent of the total mass of a cell is water. Not surprisingly, therefore, water is of prime importance for life, at any rate for those life forms which currently inhabit the Earth. All the chemical reactions of biology take place in solutions of water, which is a solvent, reactant or reaction product for biological chemistry. There are also other inorganic materials which are of basic importance. For instance, we need nitrogen for amino acids and proteins, and phosphorus for the nucleic acids. We also need sodium, potassium and other trace substances, but only in small quantities. Nitrogen, phosphorus and the metals are taken from minerals in the soil. Table 5.1 shows the distribution of various atoms in living systems compared with their distributions in the universe and on the Earth's surface.

Table 5.1 Relative abundance by mass of elements
(adapted from M. Calvin, *Chemical Evolution*, Oxford, 1969)

Element	Cosmos: Relative abundances	Earth		Life	
		Atmosphere hydrosphere	*Crust*	*Plant (%)*	*Animal (%)*
Hydrogen	1000.0	2.0	0.03	10.0	10.0
Helium	278				
Oxygen	11.8	9.978	0.623	79.0	65.0
Carbon	4.5	0.0001	0.0005	3.0	18.0
Neon	2.2				
Nitrogen	1.6	0.003		0.28	3.0
Magnesium	0.81		0.018	0.8	0.05
Silicon	0.88		0.211	0.12	
Iron	1.46		0.019	0.02	0.0004
Argon	0.14				
Sulphur	0.50	0.0005		0.01	0.25
Aluminium	0.72		0.064		
Calcium	0.09		0.019	0.12	2.0
Sodium	0.04	0.0008	0.026	0.03	0.15
Nickel	0.09				
Phosphorus	0.0094			0.05	1.0
Potassium	0.0052			0.32	0.35
Others	< 0.003	0.011	0.020	0.04	0.156

Apart from oxygen in water, carbon is the most abundant element in living systems. Often this element is in the form of complicated structures with units arranged in the shape of rings. At the present time, terrestrial living systems construct these carbon rings starting from carbon dioxide in the atmosphere. As we shall see, however, the first generation of prebiotic molecules imported from outer space almost certainly possessed these ring structures already. Life therefore has the effect of maintaining an almost constant level for at least some of these ring molecules.

Perhaps the most fundamental chemical reaction in all terrestrial life processes involves the conversion of water and carbon dioxide into exceedingly long chains of molecules known as the polysaccharides.

Starch and cellulose are the most common polysaccharides in nature and each is made up of units of sugar molecules known as glucose. The basic reaction is as follows:

Carbon dioxide (6 CO_2) + Water (6 H_2O) + Sunlight $\rightarrow$ Glucose ($C_6H_{12}O_6$) + Oxygen (6 O_2)

This reaction, which takes place in plant cells in the presence of the green substance chlorophyll, provides the nutritional source for all higher animals which live by eating plants. Furthermore, this same reaction provides oxygen for breathing. A part of the oxygen supplied by plants is converted into ozone (O_3) in the upper regions of the atmosphere by the action of solar ultraviolet light. This ozone layer has the effect of filtering out much of the solar ultraviolet radiation which would harm higher life forms. Indeed, had it not been for this filtering out of ultraviolet light, it is unlikely that the familiar higher life forms could have evolved.

Glucose molecules formed in plants by the reaction described here have the property of linking together to form the polysaccharides starch and cellulose. The glucose molecule can form two types of chain structure which differ only in the way the individual glucose units are linked together. In the two structures – cellulose and starch – the two forms of molecule can be packed into long strands of repeated units, strands which are known as polymers. The strands in starch consist of quite loosely stacked molecules compared with those in cellulose. This makes a difference in that we can digest starch but not cellulose. It is worth noting, however, that some of the primitive bacteria which inhabit the guts of horses, sheep and cattle have enzymes with the property of unzipping cellulose strands into separate glucose molecules. These animals can therefore digest cellulose in grass, while certain types of small insect can break up other forms of cellulose, as when termites feed on wood.

The formation of a polysaccharide chain by the joining together of glucose molecules involves the removal of a hydrogen atom (H) from one glucose molecule and a hydroxyl radical (OH) from the adjacent one to liberate a molecule of water (H_2O). This type of process involving the release of a water molecule in bond formation is common throughout biology, and we shall find other instances.

Cellulose forms a major component of cell walls in plants and so makes up about one third of all vegetable matter. Wood is about fifty per cent cellulose, and the conversion of wood cellulose into coal (mainly hydrocarbons) occurs when tree trunks are buried underground over extended periods of time. The overwhelming dominance of cellulose in terrestrial biology is of particular relevance to our theory of how life began. We shall see in later chapters that cellulose is not merely the most abundant organic molecule on Earth. It appears also to be by far the most abundant organic substance throughout the galaxy, and a large fraction of all the available carbon and oxygen seems to be tied up in the form of this material.

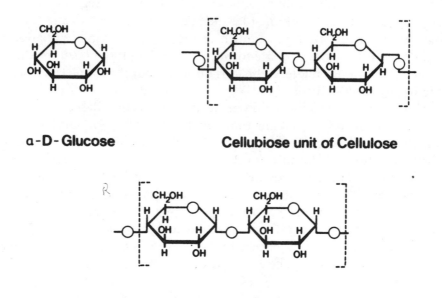

a-D-Glucose   Cellubiose unit of Cellulose

Maltose unit of Starch

Fig. 3 Structures of D-glucose and of the repeating units of cellulose and starch

Apart from the two polysaccharides derived from glucose, there are others which are also important in biology. The sugar in cellulose and starch is based on a six-membered ring molecule with five carbon atoms and one oxygen atom and known as a pyran ring, with the formula C_5O. Their chemical structures, as well as that of glucose, are shown in Fig. 3. There are also five-, four- and three-membered rings (C_4O, C_3O, C_2O) which can make sugars, the five-membered or furan ring (C_4O) being of particular importance because it makes the sugar known as ribose, a component of the nucleic acids, which constitute the genetic substance of life. Some polysaccharides involve the replacement of the hydroxyl radical (OH) groups by amino (NH_2) groups and often occur in biology. A well-known example of this kind of polysaccharide is chitin, which is the major component of the hard shells of insects and animals, such as the shell of the crab.

Polysaccharides play a dual role in contemporary biology. They hold sugar molecules in stable units for eventual use, primarily as an energy source in animals. They also serve as skeletal structures for cell wall formation, particularly in plants.

Apart from polysaccharides, there are other basic chemicals of life which form an essential part of our story. The nucleic acids (DNA, RNA) and the proteins, for instance, make up an intimate partnership in the operation of living systems. Together they are responsible for holding as well as transmitting the 'patterns' of an organism in the processes of cell growth, division and replication. The nucleic acids in themselves are able to store and copy genetic information. They are made up of a sugar (ribose for RNA, deoxyribose for DNA), a phosphate, and four bases (adenine, thymine, guanine and cytosine). (For RNA, uracil replaces the thymine of DNA.)

DNA always exists as a double-stranded molecule. An arrangement of molecular groups in a single strand of DNA is shown thus:

— Base (A, G, T or C)   Base (A, G, T or C)

| |

— sugar — phosphate   — sugar — phosphate   —

The sugar-phosphate-sugar-phosphate . . . sequences which form a spine structure are the same for all DNA strands. The genetic coding is determined by the order of the attachments of the four bases adenine (A), thymine (T), guanine (G) and cytosine (C) to the sugar-phosphate strands. The single DNA strand shown above may also be represented by the upper half of the 'rail track' diagram in Fig. 4.

In the complementary DNA strand (the lower half of the 'rail track' structure) the bases to go with T, A, G, A cannot be arbitrarily chosen, and the following pairing rules must be strictly obeyed: thymine always pairs with adenine, cytosine always pairs with guanine. These pairing rules arise essentially from the different sizes of the base molecules and the condition that the molecular ties across the two tracks must have the same lengths, very much like ties of equal length between two railway tracks.

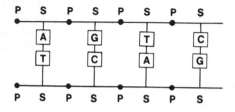

Fig. 4 'Rail-track' segment of DNA

If the 'rail track' is extended into a very long chain with each tie in the chain being one of the four permitted pairings (A-T, T-A, G-C, C-G), the pattern of ties holds all the information necessary for making an animal or plant. For a typical animal or plant the number of ties in a full strand is about a billion. In real life this long track is neither straight nor in one plane, but is twisted into a double helix and folded up so as to occupy an exceedingly small volume of space (see Fig. 5). In prokaryotic cells DNA molecules most often remain in the form of

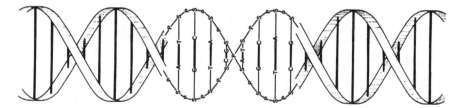

Fig. 5 Twisted-ribbon model of DNA double helix showing arrangements of molecular units. The entire structure is coiled up and occupies a very small volume of the cell nucleus

unbroken double helix coils. In eukaryotic cells, to improve efficiency of packing into the small volumes of cell nuclei, DNA strands are divided into many segments known as chromosomes, of which there are forty-six in a human cell.

RNA molecules, which are of three main types – messenger RNA (mRNA), ribosomal RNA (rRNA), and transfer RNA (tRNA) – are single-stranded molecules resembling one strand of a DNA molecule, except that the sugar in DNA, deoxyribose, is replaced with ribose, and thymine is replaced by uracil. Messenger RNA molecules have the highest turnover rate in cells and are responsible for transferring information held in DNA to the ribosomes. The ribosomal RNA together with its associated proteins interact with mRNA to order amino acids into protein chains.

An mRNA molecule can form using one strand of the DNA double helix as a mould or template. The base pairing rules in the DNA → RNA copying process are the same as those which determine the pairs within the DNA double helix which we have discussed earlier. Recently, however, it has been found that mRNA actually omits reading large sections of the nuclear DNA almost as though it has evolved in order to edit the cell's file of genetic information. This editing apart, the base ordering in the mRNA copy contains genetic information which is identical to that which occurs in the associated DNA segments which it has elected to transcribe. The information is then transferred to cell sites where protein synthesis occurs. We shall now briefly describe how this synthesis occurs.

Proteins are molecules in the form of long chains made up of units known as amino acids, of which about twenty are biologically important. The schematic structure of an amino acid may be shown thus:

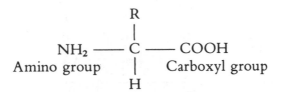

$$\begin{array}{c} \text{R} \\ | \\ \text{NH}_2 \text{———} \text{C} \text{———} \text{COOH} \\ | \\ \text{H} \end{array}$$

Amino group   Carboxyl group

The several amino acids differ only in the detailed form of the side chain designated R. The names of the twenty amino acids and their abbreviated notations are set out in Table 5.2. Proteins from widely different biological sources – for instance wheat and beef – differ only in the relative arrangements of these twenty amino acids. Proteins are like the worker bees of a living organism. They are the most active chemical ingredients of cells, taking part in myriads of biochemical reactions. These reactions are responsible for all features of cell metabolism, such as cell construction, cell replication and so on.

Table 5.2 Amino acids

Alanine (Ala)	Leucine (Leu)
Arginine (Arg)	Lysine (Lys)
Asparagine (Asn)	Methionine (Met)
Aspartate (Asp)	Phenylalanine (Phe)
Cysteine (Cys)	Proline (Pro)
Glutamate (Glu)	Serine (Ser)
Glutamine (Gln)	Threonine (Thr)
Glycine (Gly)	Tryptophan (Trp)
Histidine (His)	Tryosine (Tyr)
Isoleucine (Ile)	Valine (Val)

Table 5.3 The genetic code

1st Base \ 2nd Base	U	C	A	G	3rd Base
U	Phe	Ser	Tyr	Cys	U
	Phe	Ser	Tyr	Cys	C
	Leu	Ser	****	****	A
	Leu	Ser	****	Trp	G
C	Leu	Pro	His	Arg	U
	Leu	Pro	His	Arg	C
	Leu	Pro	Glu	Arg	A
	Leu	Pro	Glu	Arg	G
A	Ileu	Thr	Aspn	Ser	U
	Ileu	Thr	Aspn	Ser	C
	Ileu	Thr	Lys	Arg	A
	Met	Thr	Lys	Arg	G
G	Val	Ala	Asp	Gly	U
	Val	Ala	Asp	Gly	C
	Val	Ala	Glu	Gly	A
	Val	Ala	Glu	Gly	G

**** signifies chain termination

mRNA molecules carry the code for amino acid assembly to the ribosomes. Each group of three bases along the mRNA strand specifies a particular amino acid and the sequence of these triplets (typically about a thousand) dictates the sequence of amino acids in a protein. A sequence of triplets specifying a protein is called a gene and there are about a million genes in a single human cell. The prescription for defining the amino acid from base triplets is indicated in Table 5.3. This is the genetic code. By comparison with Table 5.2 we can, for instance, read that the RNA base triplet sequences UUU, CAU and GGA correspond to the amino acids Phe (phenylalanine), His (histidine) and Gly (glycine) respectively. When we consume proteins from meat or vegetables in our diet, these proteins are broken down into the constituent amino acids, and thereafter the separated amino acids are rearranged into forms which are dictated by our own DNA and which suit our own individual requirements.

Among the most important biochemicals are the 'enzymes', the family name given to a large group of proteins which make the chemical reactions in cells possible. From ancient times it has been known that yeast cells are capable of fermenting sugars by transforming sugar and water into alcohol and carbon dioxide. Louis Pasteur was the first scientist to study such processes in great detail. He observed that when glucose is stored in a sterile sealed container, no fermentation takes place. The participation of living microorganisms in fermentation was correctly inferred, but Pasteur also concluded incorrectly that intact living organisms were a prerequisite for such processes. In 1897 Edward Buchner obtained a cell-free extract by grinding yeast cells with sand and filtering the resulting material. The ability of this cell-free extract to cause fermentation led to the discovery of enzymes, and the beginnings of modern enzyme chemistry. A long struggle to isolate a pure enzyme and to study its chemical properties lasted for about thirty years. Finally, in 1926, James Summer succeeded in isolating a pure crystalline enzyme which he also showed had a protein character.

The enzymes carefully regulate the thousands of chemical reactions essential to the metabolism of living cells by speeding up, slowing down, stopping or starting the processes. There are specific enzymes for specific reactions. Some enzymes break up large nutrient molecules, for example proteins into constituent amino acids, or polysaccharides like starch into individual sugar molecules. One enzyme is a specialist at unzipping the double strands in DNA during DNA replication prior to cell division. The action of an enzyme in this process is illustrated in Fig. 6. Under the influence of enzymes the separated strands 'pull together' appropriate bases from the surrounding medium in accordance with the rules discussed earlier. In this way

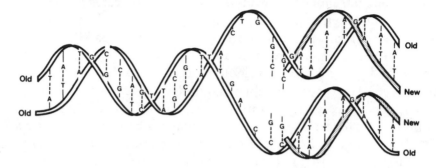

Fig. 6 The DNA double helix is split by the action of an enzyme and re-forms into two identical structures

a double-stranded DNA molecule then exactly replicates into two identical daughters (see Fig. 6). Other enzymes build proteins from amino acids, or polysaccharides from single sugar molecules. The shape of an enzyme molecule is what mainly determines its chemical function. A schematic representation can be given of the action of an enzyme in bringing together two units A, B of a substrate to produce the complex AB. The sequence of chemical events is as follows:

$$A + B + ENZYME \rightarrow A-B-ENZYME \text{ complex}$$
$$A-B-ENZYME \text{ complex} \rightarrow AB + ENZYME$$

The specificity of the enzyme – in other words its property of being only involved in certain reactions – is due to the specificity of the jig-saw type fitting of the components A, B into an active site on the enzyme molecule. An apparently simple biological phenomenon such as the transmission of a nerve impulse involves a number of steps in which one or more chemicals are converted to other substances. Enzymes play an indispensable role in all steps.

The real power house of all life on our planet is chlorophyll, which is a single type of molecule with crucial importance for the origin of

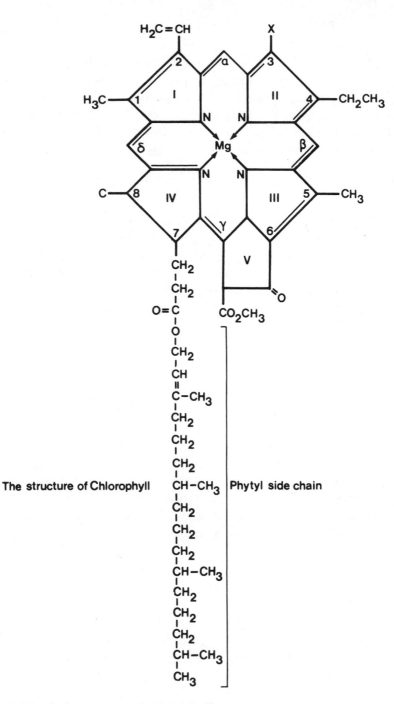

The structure of Chlorophyll

Phytyl side chain

Fig. 7 Detailed structure of chlorophyll

life. The structure of this rather complex molecule is shown in Fig. 7. Apart from the long side-chain composed mostly of hydrocarbons, the main structure consists of four five-sided rings arranged symmetrically around a central magnesium atom. Each five-sided ring is made up of four carbon atoms and one nitrogen atom. We shall later see that the ring substructures of this molecule were probably formed extraterrestrially, and probably even the first generation of entire chlorophyll molecules was imported from distant parts of the galaxy.

Chlorophyll uses the energy of sunlight to convert water and carbon dioxide into glucose, which can then be polymerized into starch, cellulose and other polysaccharides. The chlorophyll acts as a solar energy receiver and catalyst. Energy stored in the chemical bonds of polysaccharides (or of glucose) is then used to drive the complex machinery of reactions in living systems. To realize the energy stored in molecules such as glucose, a cell often uses the intermediary 'energetic molecule' ATP (adenosine triphosphate). Thus we have:

Glucose + ADP (adenosine diphosphate) + Phosphate group $\rightarrow$ ATP (adenosine triphosphate) + Carbon dioxide (CO_2) + Hydrogen (H_2)

and ATP then serves to energize further biochemical reactions.

The basic chemicals of life, many of them familiar to us after exciting discoveries in recent decades, are present in terrestrial life forms. The various substances involved – the polysaccharides, the nucleic acids, the amino acids, the enzymes and chlorophyll – are found throughout living things and they provide crucial links in our search for the origin of life.

The world of atoms | 6

Another form of evolution – the evolution of matter – has been going on in the universe all the time. Since all living as well as non-living things are made up of matter, it is an essential part of our story to look at the structure of atoms and to see how molecules and the complex chains of life have been formed. We shall then be able to understand why the universe and our own planet in particular are made up as they are – and how life has originated and evolved.

Molecules are the smallest units of a substance which still retain the chemical identity of that substance, an identity derived from the distinctive ways in which the material in question interacts with other substances or with electromagnetic radiation. Molecules, which are made up of atoms, can be small or large, depending on their nature. Hydrogen forms the smallest molecule, H_2, made up of two atomic units of hydrogen. The polysaccharides, DNA and proteins are enormous in comparison, being made up of several thousand individual atoms.

In 1911 the physicist Lord Rutherford showed convincingly that individual atoms themselves have a complex infrastructure. They are made up of a nucleus, which consists of particles known as protons and neutrons, and this is surrounded by a diffuse cloud of much lighter particles called electrons. Protons and electrons carry equal but opposite electric charges, while neutrons have no charge. The proton and neutron have about equal masses and are each nearly two thousand times heavier than the electron. Many other subatomic particles have since been discovered, but for our purposes we can still regard electrons, protons and neutrons as the principal ones.

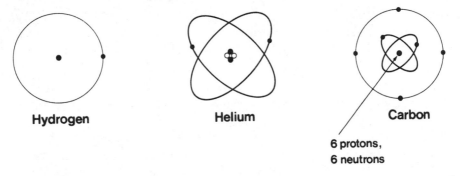

Fig. 8 Schematic models of hydrogen, helium and carbon atoms

The simplest atom, hydrogen (H), has a nucleus with just one proton surrounded by a single electron. Deuterium ('heavy' hydrogen) differs from hydrogen in that its nucleus contains an additional neutron, so that the two atoms have the same numbers of protons (in this case one) but differ in the number of associated neutrons. Two such atoms are known as isotopes. A schematic picture of the hydrogen atom is shown in Fig. 8. Helium (He) comes next in the hierarchy of atomic complexity. The common form of the helium atom has a nucleus comprised of two protons and two neutrons, with two electrons in orbit around it. In all three instances – hydrogen, deuterium and helium – the atoms are arranged so that the number of protons in the nuclei is exactly the same as the number of surrounding electrons. Since electrons and protons carry equal but opposite electrical charges, the atoms are electrically neutral. Electrons can, however, be added or removed from atoms under certain conditions, leading to the formation of charged atomic species called ions.

There are ninety-two naturally occurring elements, ranging from the simplest atom hydrogen to the largest and most complex atom uranium, which is about 238 times heavier than hydrogen (see Appendix 1). There are so far eleven more man-made, still heavier elements, making up a total of 103 elements. In atoms with more than one electron, the orbits of electrons are grouped into a succession of so-called 'shells'. The atoms themselves are arranged into a scheme known as the 'periodic table', where successive electron 'shells' are completed or filled as we progress from simple to complex atoms. The first shell (K shell) is completed with two electrons, the second (L shell) with eight, and so on in an ascending sequence to Q (see Fig. 8). In helium the K shell is completed. In carbon, which consists of a nucleus of six protons and six neutrons with six surrounding electrons, the K shell is completed and the L shell has four electrons (see Fig. 8). Since eight electrons are necessary to complete the L shell, a carbon atom has an outer electron shell with a deficit which enables the atom to share up to four electrons with other atoms, so making carbon a very reactive substance.

Atoms bind together to form molecules essentially in an attempt to complete their outer electron shells. When two hydrogen atoms come together, for instance, it is an advantage, in terms of reducing total energy, for the two single K shell electrons to be shared by both protons. Electrons are de-localized from the separate atoms, following paths around both nuclei. This leads to the formation of a chemical bond and gives a chemical reaction:

$$H + H \rightarrow H—H$$

where the line between the two atoms of hydrogen (H) denotes a shared electron for each atom. An alternative, less informative, way of representing molecular hydrogen is by the symbol H_2.

Carbon has the possibility of sharing four electrons with four hydrogen atoms to form methane; nitrogen can share up to three electrons with three hydrogen atoms to form ammonia. The structural formulae for methane and ammonia are respectively:

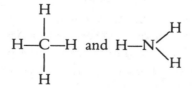

Alternatively, these molecules could be represented by the formulae CH_4, NH_3, where the explicit form of bonding is not apparent. We have already used both types of notation to describe molecules in Chapter 5. The basic rule for constructing molecules is that a single atom can have at most as many lines (valency bonds) radiating from it as the number of electrons which are required to complete its outermost shell. Thus hydrogen (H) can have only one, oxygen (O) up to two, nitrogen (N) up to three and carbon (C) up to four. The force which holds separate atoms of a molecule together arises from the sharing of outer electrons, or more correctly by the overlapping of molecular orbitals.

A hierarchy of forces holds the basic constituents of matter together: molecule to molecule in liquids and solids, atom to atom in molecules, electrons to nuclei in atoms, protons to one another and to neutrons in nuclei. The first three types of force, arising from basically the same interaction, depend directly or indirectly on the fact that electrons and protons are oppositely charged and pull towards one another. All these three types of force have similar magnitudes usually to within a factor of ten. It is these electrical forces which are involved in the melting of solids, in the break-up of molecules in chemical reactions when atoms join into new combinations with each other, and also in ionizations where electrons are stripped from atoms. The type of force involving protons and neutrons within atomic nuclei is vastly different. The nuclear forces between protons and neutrons in atomic nuclei are characterized by their enormous strengths, and by the large energy changes which they promote in nuclear reactions. Such changes occur in H-bombs, nuclear reactors, and in astrophysical processes inside stars.

Before we discuss the ways in which atomic nuclei can change their structures, let us return to look at the simplest atom, hydrogen. Here a single electron is in orbit around a proton. The size of the electron's orbit is a measure of the amount of chemical energy which is stored in the atom – the larger the orbit, the more the energy. The size of the electron orbit cannot be arbitrarily chosen, however, but must be distributed amongst a well-defined series of values determined by the laws of quantum mechanics. These sizes correspond to the various energy states (or levels) of the atom, and for hydrogen this is schematically illustrated in Fig. 9.

An electron can change its orbit from one level to another, and when this happens there is an associated absorption or emission of

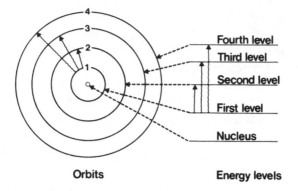

Orbits   Energy levels

Fig. 9 Energy levels of a hydrogen atom

radiation (light) at some well-defined wavelength. As the atom absorbs light quanta with greater and greater energies (shorter and shorter wavelengths), the electron orbits are pushed further out, until at a certain energy value the electron is no longer bound to the atom and is free to leave the system. This happens when light of a certain wavelength (912 Ångströms) interacts with ordinary hydrogen, or in a sufficiently hot gas with the temperature above about 10,000 degrees when collisions knock out associated electrons and the hydrogen atoms become ionized.

Similar interactions between light and atoms occur for elements other than hydrogen. Similarly too, molecules can interact with radiation. The shared electrons in molecules change their disposition by absorbing visual and ultraviolet light. Furthermore, the separate nuclei in a molecule can vibrate or wobble relative to one another, or rotate bodily, and these different types of motion are also associated with the absorption or the re-emission of radiation. Vibrational and wobbling types of motion give rise to infrared spectra which are often characteristic of a particular molecular substance. Rotational motions give rise to characteristic absorptions at microwave and radio wavelengths. We shall see later that such spectra, in ultraviolet, visual, infrared, microwave or radiowavelength regions, can give us crucial information about the presence of various types of molecule in distant parts of the galaxy.

We turn now to the question of the genesis of atomic nuclei. Just as in the case of biological species, atomic species are not fixed or immutable. The structure of individual atomic nuclei can be changed and re-shuffled under certain conditions. This happens continually in the interiors of stars, including the Sun, and it is the energy released by such nuclear processes that permits stars to shine for long periods of time. The energy we receive from the Sun comes mainly from a sequence of nuclear reactions in which four hydrogen nuclei unite to form the nucleus of a helium atom. We shall not be concerned here with the detailed nature of nuclear reactions, but only note that such reactions require temperatures of the order of tens of millions of degrees. Very high temperatures are needed to give sufficiently fast motions to the nuclei for them to come together and to fuse into larger structures. As the temperature is continually raised in an astronomical process, from tens to hundreds and ultimately to thousands of millions of degrees in a stellar interior, first we have hydrogen converted into helium, then helium to carbon, oxygen and heavier elements. The later stages of this sequential nuclear building occur when stars explode, giving rise to supernovae, a process which generates the heavier elements and also flings matter out into the space between stars. The peculiar distribution of the various chemical atoms which we see around us did not arise by any mystical accident. The fact that carbon and oxygen are so abundant in the universe and gold and uranium so rare is the result of a well-defined history (or prehistory) of nuclear evolution which occurs within stars throughout the universe.

The universe, then, possesses an amazing variety in its material make-up even on a very minute scale – in the world of particles. Matter has evolved in such a way that the universe has a very uneven distribution of substances. The arrangements and relative abundances of elements will provide important clues in the search for the origin of life.

The stars are formed | 7

The universe is certainly older than living things, but it gave rise to life and strictly conditioned it. For this reason, and because of other aspects of the theory we shall develop, the next stage is to look at the way in which the universe, with its galaxies, originated and goes on evolving. We shall see the importance of stars and the ways in which the beginning of life is bound up with them.

There have been many theories of the origin of the universe. Early ideas on this issue, just as on the origin of life, are often linked with fantasy and legend, religion and myth. The following dialogue which is alleged to have taken place between Caliph Harun-ar-Rashid (AD 766–809) and a wise man may have a relevance to modern cosmology:

> *Caliph:* I have heard that the Earth is not quite flat and I just cannot believe it.
> *Wise man:* Commander of the faithful, few men know it, but the Earth is indeed not flat. The Earth is the back of a giant turtle with sleepy eyes.
> *Caliph:* But where is this turtle? Does he stand, or does he swim?
> *Wise man:* Commander of the faithful, this is a great secret. But the turtle swims in a big sea without an end.
> *Caliph:* But where is that sea?
> *Wise man:* The answer to that question only I know in the whole world and I found it out myself. This sea is in a big ball carried on the back of a white elephant with golden tusks. But, commander of the faithful, on what the white elephant stands – no one in the world knows.

It seems that we are still groping to find where a certain white elephant stands, although the earlier steps in this logic are rather better understood.

Modern astronomy has not yet provided a final answer to the question of how the universe began. The prevalent belief is that it started with the explosion of a 'cosmic fireball' about thirteen billion years ago. This type of cosmological theory was first discussed by the Russian mathematician Alexander Friedmann (1888–1925) and developed later by a number of workers, particularly H. P. Robertson and G. Gamow. According to the theory the universe started off as a superdense, superhot mixture of neutrons and light quanta (or photons – units of light energy). In this picture a universe which is one second old has a temperature of twenty billion degrees, and after ten minutes the temperature is about five hundred million degrees. As the universe expands it becomes progressively colder.

There is evidence for a radiation bath with a temperature of three degrees which pervades space at the present epoch and which may be attributed to an original big bang. This evidence was discovered by A. A. Penzias and R. W. Wilson at the Bell Telephone Laboratories, U.S.A., in 1965. They detected radio waves at a wavelength of seven centimetres coming isotropically – in other words, equally in all directions – from the sky. Subsequently, similar signals were measured at shorter radio wavelengths and a temperature of three degrees was assigned to this background radiation. Although the existence of this radiation bath does not conclusively prove that the universe started with a big bang, it is certainly consistent with such a hypothesis.

If the universe started in this way, the synthesis of a few of the chemical elements began at a very early stage in its history. Primordial neutrons were first transformed into electrons and protons, and the build-up of nuclei into the element helium occurred when the universe remained hot and dense enough for nuclear reactions to occur. Heavy hydrogen, some lithium and perhaps a few heavier elements may also have been generated in the fireball, but the great majority of elements must owe their origin to nuclear processes in stars, as we saw in the previous chapter.

As the fireball expanded and cooled, a time came when condensations and sub-condensations separated out, to become clusters of galaxies and individual galaxies. The galaxies are still seen to be rushing away from one another, sharing in the general expansion of the universe, which presumably started with the original big bang. The question of whether the expansion will go on for ever or eventually reverse to give a contracting universe is still being vigorously debated by astronomers.

Within each galaxy there occurred further condensations which led to the formation of star clusters and of individual stars. Our own galaxy, which is one of billions of galaxies in the universe, is made up of about two hundred billion separate stars. The Andromeda nebula, which is the nearest external galaxy closely resembling our own, is about two million light years away and has a distinctively spiral structure (Plate 2). If the Andromeda nebula were viewed edge-on it would be seen to have a conspicuous central dust lane such as exists in our own galaxy. Spiral galaxies, such as our own, have a characteristic structure which is determined by young stars condensing from the gas and dust along spiral arms (Plate 3). There are, of course, other types of galaxy with different shapes and properties and some are irregular. The Large Magellanic Cloud (Plate 4), for instance, contains more gas and dust than our own galaxy and has such a high proportion of very young stars that it is thought by some astronomers to be younger than the Milky Way. By way of contrast, some galaxies are elliptical, a type that is found to occur in the Virgo cluster of galaxies (Plate 5). Such ellipticals are probably somewhat older than our galaxy, and they have no spiral arms, little or no gas, and no young stars. Plate 6 is an interesting photograph of a rich field of galaxies in a cluster including several different types of galaxy viewed in various aspects.

Stars together with gas and dust are distributed along the spiral arms of our galaxy. Fig. 10 shows a schematic diagram of our galaxy viewed face-on and edge-on. The Sun is about two-thirds of the way out from the centre to the outermost edge of the galaxy, and the radius of the galactic disc is about fifty thousand light years. The gas and dust are mostly confined to a thin central layer with a thickness of about three hundred light years. The formation of stars from this central layer of gas and dust is a continuing process. The oldest stars in our galaxy, those which are now in a much thicker disc, must have formed more than ten billion years ago. The youngest stars, on the other hand, could well be less than a million years old, so that they would have started to shine after the first men appeared on our planet.

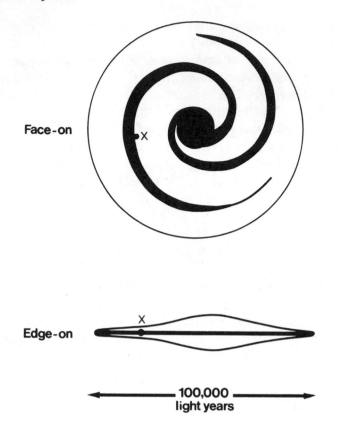

Fig. 10 Schematic views of our galaxy face-on and edge-on (the position of the Sun is shown by a cross)

Stars differ not only in their ages but also in their chemical compositions, masses, intrinsic luminosities and temperatures. The Sun has often been described as an 'average star'. It has a chemical composition which is reasonably typical of most of the stars in our galaxy as well as of the material between stars. The material of the Sun consists of about 73 per cent by mass of hydrogen, about 25 per cent helium, about 1.5 per cent carbon, nitrogen and oxygen together, and 0.5 per cent of heavier elements such as magnesium, silicon and iron. The relative abundances of the various atomic species in the solar system are set out in Appendix 1. The Sun has a mass close to two thousand million million million million grams – three hundred thousand times the mass of the Earth – and it has a surface temperature of about six thousand degrees.

Some stars are richer in metals than the Sun, while others are poorer, but in any case the variations of composition for neighbouring stars are relatively small. The majority of stars have masses which fall in the range from a few tenths of the mass of the Sun to ten times its mass. The luminosities of most stars range from less than one hundredth of that of the Sun to many thousand times as bright, while stellar temperatures range from two thousand degrees to about fifty thousand degrees. The colours of stars are related to their temperatures, the coolest appearing red and emitting their energy at wavelengths longer than seven thousand Ångströms. As we progress higher in the temperature-sequence, stars become progressively bluer, emitting their energy at shorter and shorter wavelengths. The Sun at six thousand degrees emits its peak radiation at a wavelength of five thousand Ångströms and so appears yellow in the visual. A star of fifty thousand degrees emits most of its energy in the ultraviolet.

If the luminosities of many stars are plotted against their surface temperatures the results are highly interesting (Fig. 11). Over ninety per cent of all stars fall in a band stretching across this diagram and form what is known as the 'main sequence'. The Sun belongs to the main sequence and its position is indicated in Fig. 11. This type of

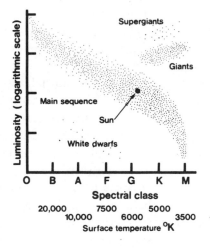

Fig. 11 H–R (Hertzsprung-Russell) diagram of stars

diagram, which was first drawn by E. Hertzsprung and H. N. Russell in 1913, is of fundamental importance for interpreting the life cycles and the evolution of stars and it is usually referred to by astronomers as a Hertzsprung-Russell (or H-R) diagram. Outside the main sequence, on which most stars are found, there are two other stellar groups. One group, lying above the main sequence, consists of the so-called giants and supergiants, which are stars with low surface temperatures but high luminosities. The second group outside and below the main sequence is made up of the white dwarfs, which are stars with comparatively hot surfaces but which are underluminous. A white dwarf has a small radius and a very high internal density.

An understanding of how stars change their positions on an H-R diagram as they evolve in time came from studies of clusters of stars. The stars of any particular cluster were formed almost at one instant from a single homogeneous cloud of gas and dust. This means that all the stars in any particular cluster would have very nearly the same age and starting composition. They would differ, however, in their masses, essentially because the separate pieces into which the cloud fragmented in the process of star formation would have a natural spread of sizes. By comparing clusters of different ages we can get 'snapshot' pictures of the progress of stars on an H-R diagram, and by combining these snapshot pictures with theoretical calculations we can trace the evolution of a star of any given mass and starting composition. Evolutionary tracks for stars of masses 1 and 10 times that of the Sun and with solar composition are shown in Fig. 12.

A star begins its life as a fragment of an interstellar gas cloud. The cloud collapses freely under the action of gravitational forces. The internal temperature and pressure rise until free collapse is slowed down by forces pushing outwards. This newborn 'protostar' derives its first energy from a slow gravitational contraction, and this energy is radiated from the protostar's surface. The protostar makes its first appearance as a low-temperature object in the H-R diagram. The central temperature continues to rise until a point is reached when the energy which the star gains from nuclear reactions taking place in its central regions comes into balance with the energy that is being lost through the emission of radiation at the surface into space. For a star of mass similar to the Sun this stage is reached after about twenty million years. The star of solar mass now arrives on the main sequence, where

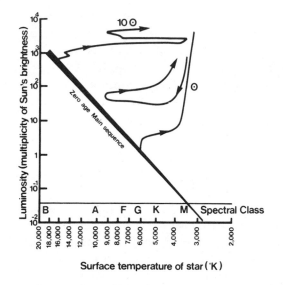

Fig. 12 Evolution of stars of 1 and 10 solar masses on an H-R diagram (⊙ = 1 solar mass; 10⊙ = 10 solar masses)

it remains for many billions of years as it continues within its interior to tap the energy derived from the conversion of hydrogen into helium. The Sun has spent nearly five billion years already in this phase, about half its total main sequence life, which is calculated to be about ten billion years.

The precise position of a star on the main sequence depends mainly upon its mass. The more massive stars have high luminosities and high surface temperatures and so lie on the top left end of the main sequence, while the less massive stars have lower luminosities and lower surface temperatures and so lie down the main sequence towards the right hand side of the H-R diagram. The evolutionary timescales also depend on the mass. The rate of consumption of nuclear fuel within a star is found to be roughly proportional to the cube of its mass. The more massive stars use up energy and burn nuclear fuel more prodigally at every stage in their evolution, with the result that they have shorter lifetimes than less massive stars. A star with a mass ten times that of the Sun, for instance, consumes its fuel about a thousand times faster and so lives only for about a hundred million years.

Stars spend over ninety per cent of their life in more or less stationary positions on the main sequence. Since the conversion of hydrogen into helium is most efficient near the centre of the star, where the temperature is highest, a central core of helium develops and begins to eat its way slowly outwards. When about ten per cent of the hydrogen in a star of solar mass is converted in this way into helium, the star changes its appearance dramatically. The helium core shrinks, and the hydrogen around it continues to be used as a major nuclear fuel. Curiously, while this is happening in the central regions, the outer regions of the star expand to such an extent that the overall radius increases by a factor of a hundred or more. Although the luminosity also increases, there is such a great expansion in size that the surface temperature actually decreases. The star appears redder, moving into the giant region of the H-R diagram, to become a red giant or supergiant. Betelgeuse in the constellation of Orion is an example of a supergiant easily visible to the naked eye. When the Sun eventually reaches the red-giant phase of its evolution in about the year AD 5 billion its radius will engulf Mercury, Venus and the Earth and vaporize them.

The switching of a star from its main-sequence position to the red-giant region is so rapid that this is probably why stars are rarely observed between these two regions (see Fig. 11). The star's red-giant phase lasts until the hydrogen in the layer outside the core is largely exhausted, perhaps a hundred million years for a star of solar mass. The helium core at the star's centre continues shrinking, getting progressively hotter as it does so until a temperature in the core is eventually reached when conversion of helium to carbon begins. The triggering off of this reaction occurs quite suddenly and the star then moves quickly away from the red-giant region. As the star evolves further, carbon is converted into still heavier chemical elements.

During the red-giant phases of their evolution, stars have low surface temperature, high luminosities, and distended stellar atmospheres. The star is also convective in the sense that matter from the interior continually bubbles up to the surface. Under these conditions material is able to escape from the surface out into space, and such mass flows have indeed been observed in many red giants. With low surface temperatures, typically of about two thousand degrees, molecules and solid particles are formed in the gas which escapes from these stars. Such mass flows from giant stars make important contributions to the molecules and dust in interstellar space.

The ultimate fate of a star, like its earlier history, also depends rather sensitively on the mass. It seems that a star with a mass comparable to that of the Sun must inevitably become a white dwarf, whereas stars more massive than about ten times the Sun eventually explode as supernovae. This is because in very massive stars nuclear reactions progress to an unstable stage. The outer layers are then suddenly and violently flung out into space. This is another important process for sporadically enriching interstellar space with heavy elements, molecules and dust.

Supernovae occur at an average rate of two or three per century per galaxy. A classic example of a supernova remnant is to be found in the Crab nebula in the constellation of Taurus (Plate 7). The supernova explosion which led to this was observed and recorded by Chinese astronomers in the year AD 1054. Matter in the form of electrons and nuclei which rushed out along the striations during this explosion is still found to be in a state of violent agitation. The residual core of the exploded star was recently found in the form of a pulsar, a form of object discovered by A. Hewish, S. J. Bell and their colleagues in 1968. The residual core of the Crab nebula emits pulsed electromagnetic radiation at all frequencies which have been observed, from radio waves to gamma rays, and the pulse period is thirty-three milliseconds. This periodicity is thought to correspond to the spin of a neutron star around its axis of rotation. Since 1968 scores of pulsars have been discovered. They all 'pulse' with almost clockwork regularity and with periods typically of about one second, significantly longer than the period of the Crab pulsar. The pulse period is thought to lengthen slowly as a pulsar ages, so that the Crab is a particularly young example, being less than a thousand years old.

The stars, then, have their own evolution and they provide a vast and variegated background against which life first started and has gone on developing. The distances between the stars are gigantic and their sizes vary enormously, from those which are considerably smaller than our own Sun to others which are many times more massive. The universe without the stars is unthinkable, but it is now equally unthinkable – as we shall see in the next chapter – without the matter that lies between them in the form of interstellar clouds.

The gas between the stars | 8

The rich display of stars on a dark night sky is a spectacle which cannot fail to impress even the most casual observer. A thick, almost white band of light, the Milky Way, consisting of several thousands of distinguishable stars, and vastly more that we cannot see separately with the naked eye, stretches all the way across the sky. Yet the total volume of space occupied by stars is an infinitesimally small fraction – less than one part in a hundred billion billion – of the volume of the galaxy as a whole. The nearest star to us, Alpha Centauri, is some 4.3 light years from the Sun, which is more or less the average distance between stars in the galaxy. The Milky Way as a whole has a thickness of several thousand light years at the point where the Sun is, and the entire disc spans a diameter of about a hundred thousand light years.

What lies in the vast recesses of space between the stars of our galaxy? Until recently, even up to the end of the second decade of the present century, astronomers were almost unanimous in thinking that between the stars there was nothing. Yet close-up pictures of star fields in the Milky Way tell quite a different story. Plate 8 shows a portion of the Milky Way in the constellation of Sagittarius. This photograph shows many dark patches and striations which we now know represent clouds of obscuring interstellar material lying in front of a more or less uniform spatial distribution of stars. This interpretation, however, did not come to be generally accepted until about the mid-1920s. For a long time astronomers had contended that the dark patches in star fields were really holes in the spatial distribution of stars.

The English astronomer W. Herschel was perhaps the first to recognize and map 'dark patches' in the sky without the aid of photography, way back in 1782. He recorded, for instance, that the Milky Way appeared to be bifurcated in the region between the constellations of Scorpius and Cygnus, and he thought he was seeing an actual recess in the Milky Way system. Herschel's observations included also the bright nebulae – hazy non-stellar wisps of light in the sky – some of which are now identified with luminous interstellar gas clouds, others with external galaxies of the type discussed in the previous chapter.

The first direct evidence pointing to the existence of interstellar gas clouds came from work by J. Hartmann in 1904. By studying the spectra of certain stars he showed that ionized sodium and calcium atoms were present somewhere along the line of sight from the Earth to these stars. Hartmann's conclusion was later confirmed by other astronomers, and by 1930 the existence of calcium and sodium ions (charged particles) in interstellar gas clouds was generally accepted. This does not of course mean that these particular ions are the primary constituents of the clouds – they are in fact present only in quantities that are small compared to other constituents, especially hydrogen – but these minor constituents played an important role in tracing and locating the clouds between stars.

The dark appearance of some interstellar clouds came to be understood in the 1920s. It became clear that interstellar clouds contained a finely divided particulate component as well as gaseous material. These interstellar dust particles have dimensions of a few ten thousandths of a millimetre, just about the size at which light begins to turn corners. Such particles are very efficient at extinguishing or dimming the light from distant stars. Comparatively small quantities of interstellar dust present in interstellar clouds could produce striking effects in causing obscuration of starlight.

It has turned out that most of the gas and dust is clumped into clouds with an average radius of about ten light years, while the average separation between neighbouring clouds is about three hundred light years. There is, however, a fairly wide spread of radii and separations about these averages, so that some clouds are much more compact, while others are much more extended. The extended clouds often appear as giant complexes with a great deal of fine structure in the form of cloudlets and fragments.

An interstellar cloud may contain anywhere from ten to about a million atoms in a cubic centimetre. Even the higher values in this density range are considerably lower than the densities attained in a laboratory vacuum system. In typical cases the gas temperature in interstellar clouds ranges from five or six degrees above absolute zero to about a hundred degrees. At such low densities and temperatures, interstellar atoms move slowly and interact with one another rather sluggishly, but the long timescales which are available – a million to a hundred million years for the lifetime of clouds – allow an interesting chemistry to develop.

The main constituent of interstellar matter is hydrogen. The bulk of interstellar hydrogen occurs as one of three forms: neutral atomic hydrogen, ionized hydrogen, molecular hydrogen. The presence of large quantities of neutral hydrogen was first detected by radio-astronomers in the 1950s. Radio emission at a wavelength of twenty-one centimetres arises when the electron in a neutral hydrogen atom reverses its spin from a direction parallel to the spin of the proton to the opposite direction. The resulting radio emission at twenty-one centimetres has provided an important probe for determining the temperature as well as the spatial distribution of neutral hydrogen. Ionized hydrogen occurs in localized regions around hot stars. Ultraviolet radiation from these stars strips off electrons from hydrogen atoms within a well-defined 'ionization zone'. Molecular hydrogen was detected about a decade ago by means of characteristic spectral lines in the ultraviolet. Since the Earth's atmosphere is opaque at ultraviolet wavelengths, these observations were made from rockets and Earth satellites above the atmosphere. (Hydrogen molecules have also been indirectly detected by radioastronomers, when they monitor the effect of collisions of hydrogen molecules with molecules of carbon monoxide.) The hydrogen molecule, H_2, is present in clouds which are dense enough to provide a screening off of the dissociative ultraviolet radiation from stars, radiation which can break the H-H bond. Hydrogen molecules form when neutral atoms of hydrogen collide on the surfaces of interstellar dust particles. A large fraction of all the hydrogen in our galaxy is in molecular form.

There are several types of motion associated with interstellar matter. Individual gas atoms have random speeds of about a kilometre per second; the clouds are internally turbulent; and they also move through space at random speeds of several kilometres per second. Shock-waves build up through the gas when newly born massive stars begin to emit floods of radiation, or when supernovae explode. Furthermore, the whole galaxy turns about its centre once every two hundred million years, with the axis of rotation at right angles to the galactic plane. Individual clouds also have their own rotary motions.

The interstellar matter is also structurally and chemically far from static. There is continual change: clouds form and disperse; cloud fragments collapse into stars; and the stars themselves throw matter back into interstellar space, in their very early life, in red-giant phases, or at the end of their evolution in cataclysmic events such as supernovae. Supernova explosions also inject extremely energetic atomic nuclei and electrons into the space between stars, so that the particles travel almost at the speed of light. These high energy particles are the so-called cosmic rays which swim around in the galaxy, being confined to staying there by magnetic fields which thread through the spiral arms.

The magnetic fields that pervade the galaxy are basically similar to force fields which are associated with electromagnets and which give rise to the familiar patterns of iron filings. Galactic magnetic fields are mostly coiled up in a helical form around spiral arms. When high-speed electrons, such as those from supernovae, are injected into these fields, electromagnetic waves – mainly at radio wavelengths – are emitted. This so-called synchrotron emission is seen to be coming diffusely from the plane of the galaxy, as well as more intensely from individual sources where rapid injections of high-energy electrons occur. Magnetic fields also show up by causing a systematic orientation of elongated dust grains, which in turn give rise to an observable polarization of visual starlight. Magnetic fields thread individual gas clouds together, as well as joining fragments which break apart from these clouds. Moreover, it is certain that such fields play an important role in controlling the dynamics of interstellar matter. Stars, including the Sun, are also known to have magnetic fields, an inheritance from their parent cloud fragments. As we shall see, stellar magnetic fields have a crucial role to play in the formation of planets.

If we look at the more tangible material content of the galaxy, we can learn something of the overall atomic constitution of interstellar matter (Fig. 13). Data are now available from various sources, including solar composition studies and meteoritic measurements, on the

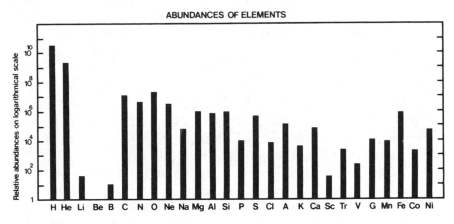

Fig. 13 Interstellar abundances of elements (for names of elements given as symbols, see Appendix 1)

average values which are appropriate to clouds of gas and dust in the vicinity of the Sun. This average atomic composition is that which would result from breaking up the molecules and dust in the clouds, and then smearing all the resulting atoms into a uniform mixture.

The molecular concentrations in the gas throughout the galaxy show wide variations which depend on physical conditions such as temperatures and densities, as well as the proximity of clouds to hot stars. Denser, cooler clouds as a rule have larger and more complex molecules, while the low-density clouds have only fairly simple molecular species – such as methylidyne (CH), cyanogen (CN), hydrogen (H_2) and carbon monoxide (CO) – in the gaseous phase. Some complex molecules in dense clouds tend to break up when stars form in their midst.

Throughout all clouds there exists the all-pervasive, yet elusive dust component. We shall return later to the story of how this dust can be identified, a problem which also turns out to relate to the composition of the more complex chemicals and biochemicals which exist in interstellar space.

It is interesting to trace the history of a single cloud of interstellar matter, one that is on the verge of collapse to form stars. The total mass of a typical interstellar cloud is several thousand times the mass of the Sun, and about two per cent of this total mass consists of interstellar dust. The gaseous component at first consists mainly of single atoms with a density of about a hundred atoms per cubic centimetre and a temperature of ten to a hundred degrees above absolute zero. When the cloud has collapsed and its density has increased tenfold to about a thousand atoms per cubic centimetre, molecules begin to be formed. The first molecules are hydrogen (H_2) and carbon monoxide (CO). Simple molecules like hydrogen can form efficiently when individual atoms stick on to dust particle surfaces, waiting there until an appropriate 'mate' arrives on the surface, and then bouncing back as molecules into the gas.

Reactions in the gas phase are slow if the atoms are all neutral, but cosmic rays which penetrate the gas cloud can ionize a small fraction of atoms and this greatly enhances the rates of molecule formation. This is because at the low densities appropriate to such clouds, chemical reactions between ions (charged atoms) and neutral atoms can proceed with much greater efficiency than reactions between purely neutral atoms. This type of chemistry, which is known as ion–molecule chemistry, was first developed in an astronomical context by P. M. Soloman and W. Klemperer in 1970.

Radioastronomers have recently discovered a wide range of molecular species in clouds which are probably at just about the stage of collapse discussed in the previous paragraph. Table 8.1 shows a list of all the molecules which have been so far observed, mainly by radioastronomical techniques. Many of the complex molecules in clouds have been identified by tuning radio receivers to precisely the frequencies at which the molecules in question are known to absorb or emit radiation. The main difficulty in this technique lies in predetermining the correct radio frequencies which characterize given molecules. Once these are known, the technical problem of finding the molecules is relatively simple. The observed molecules are therefore not a fully representative list of what is or may be there. In fact much more complicated molecules might exist in these clouds and not be detected by radioastronomers, mainly because of difficulties in predetermining the right radio frequencies.

Table 8.1

OBSERVED INTERSTELLAR MOLECULES
in dense molecular clouds

INORGANIC		ORGANIC
H_2 hydrogen OH hydroxyl SiO silicon monoxide SiS silicon sulfide NS nitrogen sulfide SO sulfur monoxide	DIATOMIC	CH methylidyne CH^+ methylidyne ion CN cyanogen CO carbon monoxide CS carbon monosulfide
H_2O water N_2H^+ H_2S hydrogen sulfide SO_2 sulfur dioxide	TRIATOMIC	CCH ethynal HCN hydrogen cyanide HNC hydrogen isocyanide HCO^+ formyl ion HCO formyl OCS carbonyl sulfide
NH_3 ammonia	4–ATOMIC	H_2CO formaldehyde HNCO isocyanic acid H_2CS thioformaldehyde
	5–ATOMIC	H_2CHN methanimine H_2NCN cyanamide HCOOH formic acid HC_3N cyanoacetylene
	6–ATOMIC	CH_3OH methanol CH_3CN cyanomethane $HCONH_2$ formamide
	7–ATOMIC	CH_3NH_2 methylamine CH_3C_2H methylacetylene $HCOCH_3$ acetaldehyde H_2CCHCN vinyl cyanide HC_5N cyanodiacetylene
	8–ATOMIC	$HCOOCH_3$ methyl formate
	9–ATOMIC	$(CH_3)_2O$ dimethyl ether C_2H_5OH ethanol HC_7N cyanotriacetylene

Even so, the list of molecules already found (Table 8.1) is quite exotic. For instance we note that formaldehyde (H_2CO), a basic molecular unit from which sugars and polysaccharides could form, is abundant and widespread in the galaxy. Furthermore, a molecule of formic acid (HCOOH) and a molecule of methanimine (H_2CHN) could react to give the simplest amino acid, glycine (NH_2CH_2COOH), and there is every reason to believe that this will happen very extensively. So a quite complex prebiotic chemistry seems to be taking place already at the stage of prestellar collapse of dense interstellar clouds. There is empirical evidence that clouds in this state are held back from falling freely under gravity and the force that is holding them back is almost certainly magnetic. The general magnetic field of the galaxy would have threaded through the collapsing cloud. The partially ionized gas in the cloud would be tied to these magnetic field lines, which would be squashed in the process of collapse. This process provides a resistive force which slows down the collapse, to a degree where condensation takes tens of millions of years to become effective.

When the collapsing cloud becomes sufficiently dense, yet still far removed from a truly prestellar state, complex molecules are deposited as tar-like sticky coatings on the dust grains. The grains themselves now collide sufficiently often with one another to enable growth by sticking together to produce larger grains ten to thirty times their original radii. Such composite dust grains contain a hybrid mixture of many organic materials. We expect a substantial proportion of these grains to survive through the subsequent evolution of the cloud and eventually to be returned with some of the gas into the interstellar medium. This could provide an important source of solid particles with a complex organic chemical composition.

The next stage is for cloud fragments – many hundreds of them – to break apart and collapse separately to form individual stars. It is difficult to determine what fraction of the total mass of the original cloud ends up as stars, but a fraction of one third would seem a likely upper limit, though it could be appreciably less than this. A large part of the gas with its complex molecules must certainly be returned into the general insterstellar medium before the molecules are destroyed by ultraviolet radiation from the newborn stars.

The Orion nebula is a spectacular example of a region of the galaxy where stars are forming at the present time. Young luminous stars, clouds containing highly complex molecules, dust and ionized gas coexist over a large volume (Plate 9). This could be a place where the birth of new stars goes hand in hand with the formation of complex biochemicals.

Biochemical substances form in high–density clouds like the Orion nebula, but only a small fraction of these molecules are engulfed by the newborn stars or suffer break-up by the radiation from hot stars. Most of the molecules are thrown back into the space between stars. An individual clump of grains containing organic molecules can be shuffled back and forth between the denser star-forming regions and the less dense interstellar clouds very many times, before being finally destroyed by direct incorporation into a star. During this long process of movement to and fro between widely different physical conditions in different regions, it seems inevitable that a Darwinian–style molecular evolution will occur. This would probably lead to a widespread emergence of those molecular structures which can best withstand the most hostile conditions encountered during this whole process of cosmic turmoil. A molecular species strongly resistant to break-up by the various disruptive agencies in interstellar space would finally emerge as being most widespread in the galaxy.

It is therefore only a part of the gas in space that goes towards the formation of stars, the rest of the gas being subject to a complex process of prebiotic chemistry in which the basic materials of life – the biochemicals – are formed.

From interstellar gas to dust | 9

The dust between the stars is important for the whole evolution of the universe and for the origin of living things. What is the precise chemical composition of these interstellar dust grains? How and where do they originate? What is their role in the formation of stars and planets? Do they have a relevance to the origin of life on our planet? These questions have presented a continuing challenge to scientists for over four decades. Although no final answers have been yet agreed upon, vital clues which may lead to a solution of some of these questions are at last becoming available. We shall describe the new evidence and the conclusions that can be drawn from it in the present chapter and in the one which follows.

Interstellar dust is visually the most conspicuous component of the matter between stars. It shows up as dark patches and striations against the background of distant star fields. The light from individual stars is dimmed as well as reddened by its passage through interstellar clouds, where it interacts with dust particles. The reddening occurs by a process similar to the reddening of a street lamp through a fog, or the reddening of sunlight in the evening sky. Blue light is scattered as well as absorbed more strongly by dust or molecules than is red light. In the case of natural light, which is a mixture of all colours, blue light is preferentially filtered out while the red light is transmitted. The result is that an original light source – Sun, star or street lamp – appears reddened.

Dust makes up one or two per cent of the total mass of the interstellar clouds. Since we know that the overall atomic constitution of interstellar matter is such that carbon, oxygen and nitrogen together comprise a very similar fraction of the total mass, it is likely that these elements, or combinations of them with hydrogen, are assembled in dust grains. More abundant elements such as hydrogen and helium cannot form solids under interstellar conditions, while elements like magnesium, silicon and iron are too low in abundance in the interstellar medium, by a factor of at least three, to make up the major component of dust.

Early investigations of interstellar dust were mainly restricted to a study of the way the dust absorbs and scatters starlight. The first attempts to obtain quantitative estimates of this dimming – or extinction, as it is called – of starlight were made in the 1930s. It was shown that at a single photographic wavelength close to 4500 Ångströms the dimming of starlight amounted to a reduction of intensity by a factor of about two for every three thousand light-years of passage of the light through interstellar space. From this information it was easy to infer that interstellar dimming could only be reasonably attributed to solid particles (or molecular aggregates) which have dimensions comparable with the wavelength of light. Other kinds of absorbers, for example free electrons, separate atoms or small molecules, could be shown to need implausibly large density values if they were to produce the observed amount of dimming.

With the advent of photographic filters and photoelectric techniques in astronomy it was possible to study quantitatively how interstellar dimming varies with wavelength. This is done by comparing the spectra of two stars which are intrinsically similar, one of which is more dimmed than the other. Such comparisons provide information on the wavelength dependence of extinction caused by interstellar dust. The relationship between extinction and the wavelength – the so-called extinction curve – provides an important item of information which has a bearing on the properties of interstellar dust grains.

The first extinction curves for starlight encompassing the wavelength region 5500–3500 Ångströms were obtained by J. Stebbins, C. M. Huffer and A. E. Whitford in the 1930s. Over this particular wavelength region it was found that the opacity of interstellar dust was proportional to the reciprocal of the wavelength – in other words, when the wavelength doubled, the opacity was halved. And precisely the same type of relationship was found to hold over wide

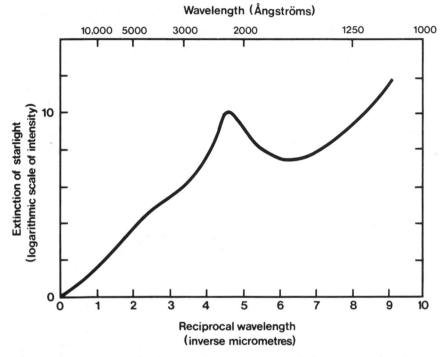

Fig. 14 Dimming – also known as extinction or absorption – of starlight in its passage through cosmic dust clouds

areas of the sky. In recent years the wavelength range of the interstellar extinction curve has been progressively stretched further into the ultraviolet (shorter waves) as well as into the infrared (longer waves). The latest data on the extinction of starlight by interstellar dust are shown in the form of an 'average' extinction curve in Fig. 14. Data in the ultraviolet, at wavelengths shorter than 3000 Ångströms, were obtained by T. P. Stecher, R. C. Bless and A. D. Code using equipment carried on rockets and Earth satellites. The most conspicuous feature in this extinction curve is a broad hump centred on the wavelength 2200 Ångströms in the ultraviolet.

Up to 1964 the interstellar extinction curve was only available for infrared and visual wavelengths longer than 3300 Ångströms. Over this limited waveband there were numerous attempts to fit observations with various types of dust grain model. From 1942 to 1964 the most widely accepted view was that interstellar grains consisted of a mixture of frozen ices – water-ice, solid methane and ammonia-ice. Grains consisting of this material fitted all the available data on the extinction and scattering of starlight so well that this particular model was accepted almost as an article of faith by astronomers. The ice grain theory was developed in the 1940s by the Dutch astronomers H. C. van de Hulst, D. ter Haar and others following an earlier suggestion by B. Lindblad in 1935 that interstellar dust condenses out of tenuous interstellar gas clouds. According to the theory, the first step in this condensation process involves the formation of molecules comprised of two atoms (diatomic molecules) such as methylidyne (CH) from the interstellar gas; then these smaller molecules add on extra atoms to form larger structures. Under the tenuous conditions prevailing in interstellar clouds, however, two-atom reactions leading to the production of diatomic molecules proceed very slowly, far too slowly in fact to provide an effective route to grain formation out of single atoms.

Faced with this almost insurmountable difficulty of forming ice grains in interstellar space, we ourselves proposed a different approach to the problem of dust grain condensation in 1962. We argued that solid carbon in the form of graphite may be the main component of interstellar dust, and that such particles are formed not in interstellar space but in the atmospheres of a class of red-giant stars known as carbon stars. Carbon stars are known to have surface temperatures which vary from 1800 to 2500 degrees over a period of about a year, and they are known to possess more atmospheric carbon than oxygen. We showed that under these conditions graphite particles with radii of a few hundred Ångströms form and are expelled into interstellar space, the expulsion being caused by the pressure of light from the parent star impinging on the surfaces of the condensed grains. There is observational evidence which points strongly to the existence of dust around carbon stars. The variable and highly luminous carbon star R Corona Borealis is a somewhat spectacular example. Here we see direct evidence of a star erratically puffing out clouds of carbon 'soot' into the space between stars. More recently, astronomers have detected thermal heat radiation from carbon stars which is consistent with the presence of graphite particles.

Unfortunately, solid carbon in the form of graphite does not have any characteristic spectral signatures in the infrared or visual spectral regions, so that a definite identification of this material has proved difficult. In the ultraviolet, however, the observed hump in the inter-stellar extinction curve at 2200 Ångströms has an uncanny resemblance to an absorption feature at this wavelength which we calculated for spherical graphite particles in 1965. It seemed natural to attribute this interstellar feature to graphite and many astronomers still believe that this is valid. We now think, however, that the story is not so simple. Calculations show that the graphite particle extinction feature is centred precisely on the desired wavelength (2200 Ångströms) only if the particles are sufficiently small (less than 300 Ångströms in radius) and have almost perfectly spherical shapes. In the real astronomical situation we would expect graphite particles to have a wide range of sizes as well as shapes. While a presumed range of sizes and shapes would certainly not eliminate the 2200 Ångström feature, the feature would be made too broad and shallow to fit the detailed shape of the astronomical band. We now believe that the observed astronomical feature is due to a molecular absorption in complex organic molecules superimposed upon the broader absorption band arising from a distribution of graphite grains of various shapes. We shall return to this point, as well as to an identification of the organic molecules, in the next chapter.

Observations by R. E. Danielson, N. J. Woolf and J. E. Gaustad in 1965 already seemed to exclude water-ice as being a major component of the dust between stars. These astronomers were the first to search for a characteristic absorption feature of water-ice at the infrared wavelength 3.1 microns in spectra of highly dimmed stars. The lack of this feature in the spectra of several stars led to the conclusion that ice particles, if they exist at all, can make at most a very minor contribution to the dimming of visual starlight. Although 3-micron absorption bands have been subsequently detected in a number of astronomical sources, these most probably arise from matter local to the sources themselves and not from dust in the general interstellar medium. We shall show in the next chapter that even within these sources, the assignment of the 3-micron band to water-ice is unlikely to be correct. Convincing evidence that dust grains in many regions of the galaxy cannot be composed of water-ice also comes from several observations which show that heated dust clouds are emitting infrared radiation at temperatures which are well above the boiling point of water.

Solid carbon in the form of graphite was the first non-volatile material to be seriously considered by astronomers as a component of interstellar dust. Particles of this type could withstand very high temperatures, up to 2500 degrees above absolute zero. They could emit high frequency infrared radiation when dust clouds are located close to hot stars. Under conditions prevalent in normal interstellar clouds, carbon grains would serve as catalysts for the formation of interstellar molecules, particularly interstellar hydrogen (H_2).

Another important astronomical observation which is attributed to the properties of dust is the phenomenon of interstellar polarization. The light from distant stars is observed to be partially plane-polarized; in other words, the light waves show a preference for a particular direction of vibration. To produce the observed polarization, interstellar dust particles must be assumed to have highly elongated shapes – something like needles – and to be systematically aligned over considerable volumes within the galaxy. A large body of observational data is now available relating to the amount, direction and wavelength dependence of interstellar polarization. The observed directions of polarization in different parts of the sky imply that the long axes of dust grains are aligned preferentially perpendicular to the direction of the galactic magnetic field. This feature strongly supports the theory that the galactic magnetic field is directly or indirectly responsible for the systematic orientation of needle-shaped interstellar dust. The observed wavelength dependence of the polarization is not consistent with extinction by orientated particles composed of metal or of graphite. A non-metallic type of grain is required to explain it.

What, then, are we to make of interstellar dust? How did its grains play a part in the origin of life? It is clear that not all the available data on the scattering, extinction and polarization properties of dust grains can be accounted for by graphite particles alone. Graphite has strongly metallic properties, which means that its finely divided particles absorb much more than they scatter visible light. A layer of finely divided graphite particles would thus have a very low reflectivity (albedo) for visual light and would look black. Interstellar dust, on the other hand, is known to have a high albedo for visual light. This fact, together with other observational data, including polarization measurements, points to a mixture of graphite particles with a less strongly absorbing type of grain species. Recent ideas on the nature of this non-absorbing grain material have been varied – inorganic ices, organic polymers and silicates have been put forward as alternative possibilities – but in the next chapter we shall show that the evidence for the presence of organic prebiotic polymers in interstellar dust is very strong.

The beginnings of biochemistry | 10

Our search for the origins of life brings us to the materials of the interstellar dust clouds. These materials emit and absorb radiation at ultraviolet, visual and infrared wavelengths. By studying the emission and absorption patterns, particularly for the dust, and by comparing them with patterns found for substances in the laboratory, we can hope to determine the chemical nature of the interstellar material itself.

A difficulty intervenes, however. The way in which a material system emits radiation depends also on its temperature and on the quantity of material comprising the system. When the quantity of material is large, radiation emerges only after a multitude of emissions and absorptions within the system, and this has the effect of hiding distinctive properties of individual atoms and molecules. The chemical nature of the material becomes suppressed, with the emerging radiation being dependent only on the temperature. The radiation is then said to be of the 'black body' type.

This same difficulty confronted astronomers a century ago, when they first attempted to determine the chemical nature of the atoms in stars. To a first approximation the radiation from stars is of the black body type, as is shown in Fig. 15. In an earlier chapter we noted that the surface temperatures of stars range from about 2000 degrees for the coolest stars up to about 50,000 degrees for the hottest, and Fig. 15 shows a number of examples in this range. For actual stars the smooth curves of Fig. 15 are found to contain ripples, and astronomers learned to look for the ripples since it was these which contained the sought-for chemical information.

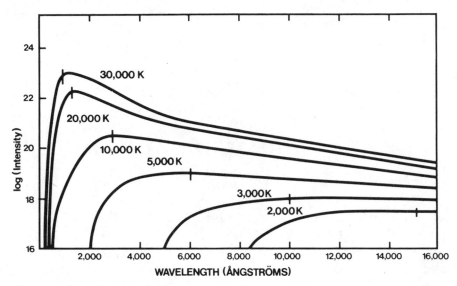

Fig. 15 Black body spectra for various temperatures (vertical marks show wavelengths of peak emission)

In a similar way we may expect to observe radiation approximately of a black body type coming from the interstellar dust clouds, but with deviations from strictly smooth black body curves that contain the chemical information we are seeking. We shall refer to such deviations as 'features'. Temperatures are much lower for the interstellar clouds than they are for stars, and temperatures are also lower for the shells of dust which sometimes surround stars. Thus temperatures in the range from 2000 degrees down to low values, say 20 degrees, are associated with the emission of infrared radiation by circumstellar and interstellar clouds. Over this temperature range the peak emission of black body radiation goes from about 1.5 microns at 2000 degrees to 150 microns at 20 degrees, encompassing the 'near-infrared' at one extreme and the 'far-infrared' at the other.

The birth of infrared astronomy a little over a decade ago brought a spate of important new discoveries. Advances in the techniques of infrared detection deploying a variety of solid-state low-temperature devices have made several infrared wavebands accessible to measurement. In some cases it is possible to 'see' infrared sources with appropriately located ground-based telescopes using natural 'windows' in the Earth's atmosphere, a window being a waveband for which the atmosphere is adequately transparent. In other cases high-altitude aircraft and balloons have been used.

The earliest sky survey in the infrared was carried out about ten years ago by R. B. Leighton and G. Neugebauer at the single wavelength of 2.2 microns, and an impressive list of over five thousand bright sources was found. Many of these near-infrared sources were not readily associated with any visible star and are believed to be 'protostars' of the type we have discussed in an earlier chapter. They are clouds of gas and dust collapsing towards a state of becoming stars. The infrared radiation which is detected could be derived in some cases from the energy of gravitational collapse before nuclear reactions have switched on. In other cases we are most probably detecting radiation from cocoons of molecules and dust which must inevitably envelop a new-born star. The dust shell absorbs the visual and ultraviolet radiation emitted by the star, thereby extinguishing it almost completely. The radiation thus absorbed has the effect of heating the dust grains to temperatures of several hundred degrees, causing them to emit infrared radiation.

Subsequent infrared studies of astronomical objects extended these measurements to other wavelengths. Perhaps the most dramatic of the early discoveries in infrared astronomy was made by E. P. Ney, D. A. Allen, W. A. Stein, J. E. Gaustad, F. C. Gillett and R. F. Knacke in 1969. They found a broad spectral feature in the 8–12 micron waveband appearing in emission in the spectra of several oxygen-rich supergiant stars. These are stars in the red-giant region of the H–R diagram which have an excess of oxygen over carbon, and with surface temperatures close to 3500 degrees. They are stars which are known to be losing gas from their surfaces. What was observed in the infrared spectra appeared to be separable into two distinct components. The star itself had its own contribution to radiation at the shorter infrared wavelengths. In addition there was a longer wavelength emission which corresponded to a temperature of less than 1000

degrees, and superposed on this spectrum there appeared a conspicuous non-black body feature in the 8–12 micron waveband. The situation is shown in Fig. 16. A very similar emission feature was also
detected in the infrared spectrum of the Trapezium nebula, a cloud of

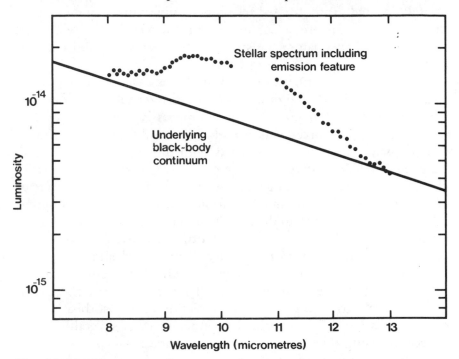

Fig. 16 10 micron excess in spectra of cool oxygen-rich stars

gas and dust in the constellation of Orion. In these sources it is clear
that we are witnessing a distinctive spectral signature corresponding
to some solid or molecular species heated to temperatures of several
hundreds of degrees above absolute zero. The hunt for this material,
which we now believe holds a clue to the origins of life, has been one
of the more exciting pursuits of modern astronomy.

The earliest theory was that this 8–12 micron feature was diagnostic
of the presence of solid particles comprised of mineral silicates –
material similar to that which makes up ordinary terrestrial rocks. It
was argued that such solids may condense out of gaseous material
flowing out of oxygen-rich supergiant stars. Furthermore, it was
suggested that these silicates could provide the major missing dielec-

tric component of interstellar dust to which we referred in the previous chapter. A difficulty with this suggestion, however, is that the cosmic abundances of magnesium and silicon seem inadequate to give a sufficient quantity of silicates to match the observed amount of interstellar dust. While we do not doubt that some silicate particles exist, it would seem better to make up the bulk of the interstellar dust from the cosmically more abundant atoms of carbon, nitrogen and oxygen.

Another difficulty for the silicate hypothesis is that with improving equipment other features outside the 8–12 micron wavelength have now been discovered. In particular, several sources of infrared radiation have been found with a broad absorption band centred at about 3 microns. The latter feature has been attributed to water-ice particles, with a variable ratio of water-ice to silicates being required to explain the properties of different sources. According to this theory, therefore, complex and variable mixtures of ice, silicates and a further still-unidentified material responsible for the overall black body radiation would be needed.

Early in 1977 we ourselves became convinced that it would be far better if a single chemical substance could be found to explain all the main features of the infrared radiation from astronomical sources. What single substance, we asked, has absorptions in the 2–4 micron band, the 8–12 micron band, and also a feature centred on 18 microns in the deeper infrared? What substance could also serve to explain the trend toward black body emission over the remaining portions of the total wavelength range? We felt that the answer to these questions must turn on an organic or biochemical material, since the abundance problem mentioned above requires that the sought-for material should be composed predominantly of atoms of carbon, nitrogen and oxygen.

The first possibility we explored, together with J. Brooks and G. Shaw, was that it might be a rather exotic material, sporopollenin. This substance forms a major component of the protective coatings in pollens and spores and is chemically and thermally extremely stable. The laboratory measures for sporopollenin gave qualitatively the required properties. There were absorption bands at 3 microns, an 8–12 micron absorption, and also a band at 18 microns – just the wavelengths where astronomical sources show characteristic features. However, a further very large absorption at 3.4 microns, due to C–H bonding, was clearly unsatisfactory and this led us to look at other biochemical materials.

It was only then, somewhat belatedly, that we asked ourselves a crucial question: What are the infrared properties of the *most abundant* terrestrial organic substance, cellulose? A dash to the library, and we found to our amazement that laboratory measurements for cellulose over the wavelength range from 2 to 30 microns showed just the absorption bands we were seeking. Moreover, cellulose was free of unwanted bands. It was only a half-hour's calculation to verify that cellulose can account both for the 8–12 micron emission feature and for the 18 micron band in the observed spectrum of the Trapezium nebula (Fig. 17). This close agreement – closer than seemed possible

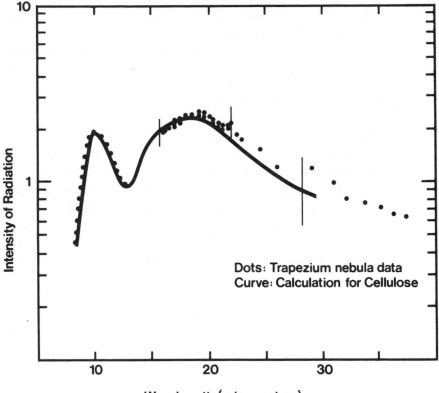

Fig. 17 Thermal radiation from Trapezium nebula compared with calculations for cellulose particles heated to 175 degrees

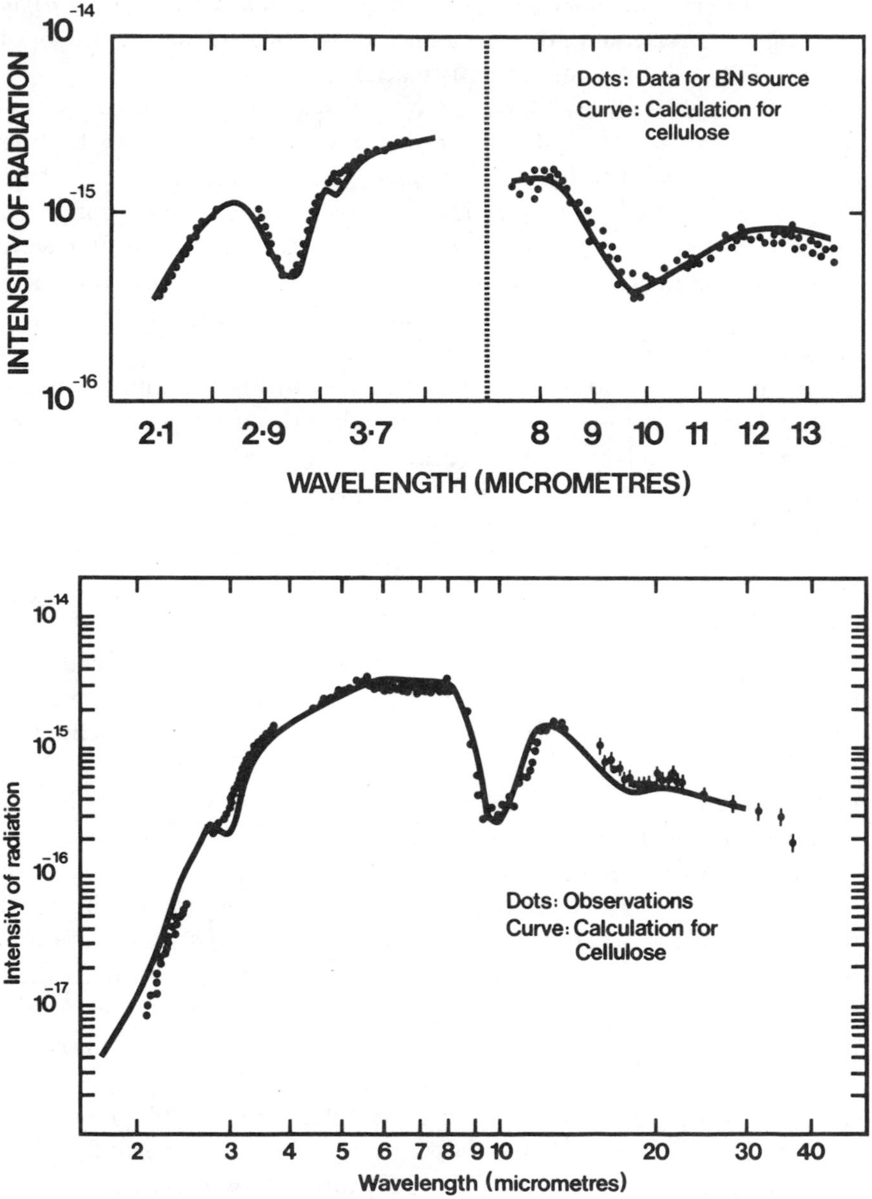

Fig. 18 Agreement between observed infrared radiation from astronomical sources and cellulose models (BN = Becklin Neugebauer)

for any mineral substance – convinced us that there was a strong *prima facie* case for saying that interstellar dust consists in the main of cellulose, or of some related polysaccharide.

Together with A. H. Olavesen we subsequently obtained laboratory data for a number of other polysaccharides, eventually finding that an ensemble of polysaccharides together with a small admixture of simple hydrocarbons provides a close fit to a wide range of observed astronomical infrared sources (see Fig. 18). No silicates were required, nor an *ad hoc* assumption of variable amounts of water-ice, nor any further unidentified material to provide a black body type emission. The differences of the observed sources (Fig. 18) arise only from temperature and structural differences within the sources themselves. There is always the same polysaccharide combination in every case, with identical infrared properties. The agreement between calculation and observation seemed close enough to justify the conclusion that polysaccharides with some hydrocarbons are present in large quantities within the interstellar clouds.

The cellulose strand is a complex structure, and one can wonder how a giant molecule of such a highly organized form could be present in interstellar space. Our first idea for explaining the origin of cellulose and other polysaccharides depended on the alternations and compressions of the interstellar gas that were discussed in the previous chapter. It will be recalled that compressions led to molecule formation, while expansions led to the break-up of the more fragile types of molecule. Alternations of these opposing effects seemed to offer scope for an exotic kind of molecular evolution to proceed.

For example, gaseous formaldehyde (H_2CO) is one of the most ubiquitous of the organic molecules actually found by observation in interstellar space. When gas clouds undergo compression formaldehyde molecules condense in solid form onto dust grains – for instance onto graphite particles – which exist already. Following condensation into solid form, a process of polymerization involving the linking-up of formaldehyde molecules could take place. The Russian biochemist V. I. Goldanskii has argued that formaldehyde polymerization would indeed be rapidly induced by the penetration of high-energy cosmic ray particles into the solid formaldehyde. The most stable form of polymerization would be into polysaccharides. Six formaldehyde molecules can be made to form a ring substructure, and a number of rings can then be linked to form polysaccharides, in particular cellulose and starch.

There is some similarity in this polysaccharide formation process to a classic experiment carried out by the Russian chemist A. Butlerov in 1861. Butlerov showed that the effect of adding an alkali to a water solution of formaldehyde was to produce a mixture of sugars. More recently, in 1965 the Ceylonese chemist C. Ponnamperuma showed that sugars and polysaccharides form as a result of shining ultraviolet light on formaldehyde.

But these processes require formaldehyde to have been already there in the first place, and one can ask how this came to be so. In attempting to answer this further question we came to a still more radical conclusion, namely that the site of origin of most if not all of the interstellar organic materials could well be in the outflows of gaseous material from the surfaces of stars. To our surprise we found that a remarkable form of prebiologic chemistry can develop in the case of outflows from hot, young, highly luminous stars, outflows at speeds of about a hundred kilometres per second. The highly energetic light quanta from the central star are systematically degraded by absorptions and emissions as the light penetrates through the outflowing material, but even so the quanta remain energetic enough to keep carbon and oxygen largely in atomic form, not as molecules of carbon monoxide (CO). At first sight, keeping the carbon and oxygen mostly in atomic form would not seem at all favourable to the building of highly complex molecules within the outflowing material, and indeed there would be little in the way of such molecules produced if they were built painstakingly atom by atom.

To illustrate how a quite different form of origin for complex molecules is possible, we can consider the formation of water drops within a cloud of the terrestrial atmosphere. Our first idea of how a new droplet forms would be through a progressive addition of water (H_2O) molecules: first two such molecules, then three, then four, and so on. But a quite different way is to begin with an already existing drop which divides into two drops. Each of the two drops then grows and divides, now to make four drops. Each of the four drops grows and divides, to make eight, and so on until a whole rain cloud is produced. Obviously we still have to make the first drop through a painstaking beginning – two molecules, three molecules . . . – but such a tedious operation is required only once, not many times over. In short, the first drop acts as a kind of template for producing the others.

The same idea can be applied to polysaccharide formation once the temperature within gas flows from stars falls below about 1200 degrees. The sequence of events leading to the formation of polysaccharide chains is shown schematically in Fig. 19. The process turns out to have an almost explosive quality about it. A single initial short

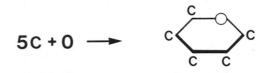

$$5C + O \longrightarrow$$

Five C atoms and one O link to make a pyran ring C_5O

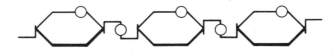

Pyran rings then link via O atoms to form skeleton structure of a polysaccharide

Fig. 19 Formation of pyran rings by linkages of carbon and oxygen atoms, and the linkage of pyran rings into polysaccharide skeletons

polysaccharide chain can produce an enormous number (10^{30}) of much longer daughter chains, due to a rapid growth of length for each chain interspersed by repeated chain fragmentations. There is an uncanny similarity to processes of biological replication (for example, cell division), where molecules grow and break up, replicating the same organizational structures from a surrounding nutrient medium. It is an interesting thought that biology learned some of its most fundamental properties and traits already at this very early stage.

It is well known among astronomers that a critical difference in the oxygen to nitrogen balance exists in stellar mass flows, depending on whether the outflowing material has experienced nuclear reactions of

Plate 1 Section of carbonaceous chondrite Mokoia, which fell in New Zealand on 26 November 1908. The spherical inclusions are chondrules which probably formed within the solar nebula. The dark matrix contains fine-grained material possibly in part of interstellar origin (Courtesy The Smithsonian Institution)

Plate 2 Great spiral galaxy in Andromeda with two satellite galaxies (Courtesy Hale Observatories)

Plate 3 Spiral galaxy in Virgo seen edge-on. The horizontal dark ring is caused by absorption (Courtesy Hale Observatories)

Plate 4 An irregular galaxy: the Large Magellanic Cloud (Courtesy Mount Stromlo Observatory)

Plate 5 An elliptical galaxy, which is a radio source, in the Virgo cluster (Courtesy Hale Observatories)

Plate 6 A cluster of galaxies in the constellation Pavo. This rich cluster includes examples of a barred spiral galaxy, elliptical galaxies, edge-on spiral galaxies, and possibly some interacting galaxies (Courtesy UK Schmidt Telescope Unit of the Royal Observatory, Edinburgh)

Plate 7 The Crab nebula in Taurus: the nebula resulted from a supernova explosion in 1054 (Courtesy Hale Observatories)

Plate 8 The Milky Way in Sagittarius showing dust clouds (Courtesy Lick Observatory)

Plate 9 The Orion nebula, a region of active star formation, with dense clouds containing complex organic molecules (Courtesy Lick Observatory)

Plate 10 Comet Kohoutek: several organic molecules were discovered in the coma and tail (Courtesy Joint Observatory for Cometary Research operated by NASA/Goddard Space Flight Center and New Mexico Institute of Mining and Technology)

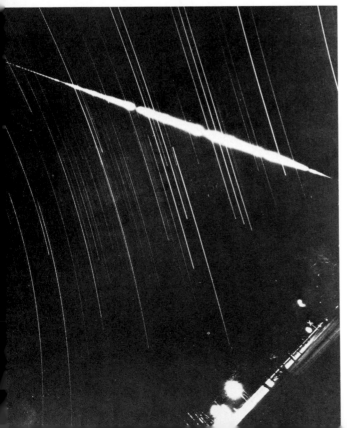

Plate 11 A meteor trail: it is caused by the burning up of surface layers of a meteoroid as it rushes through the Earth's atmosphere. The trail shown lasted 9 seconds and the meteorite which landed weighed 22 lbs (Courtesy Prairie Meteorite Photography and Recovery Network, Smithsonian Astrophysical Observatory)

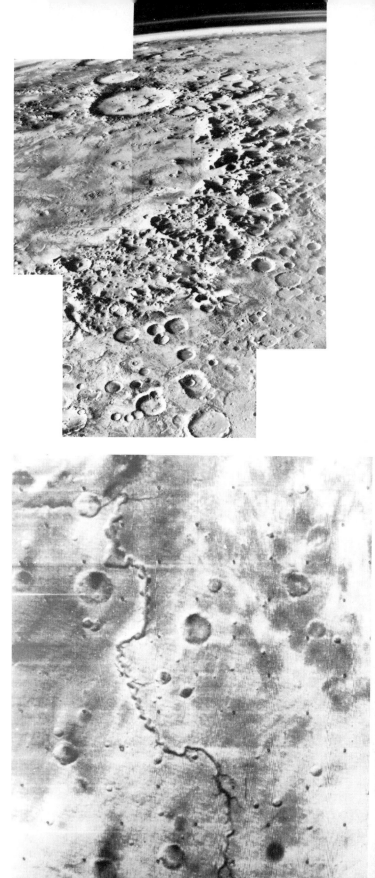

Plate 12 Close-up of Mars, across the Argyre Planitia: photograph taken 11,200 miles away by one of Viking Orbiter 1's two television cameras on 11 July 1976 (Courtesy NASA)

Plate 13 Landscape of Mars showing valley with tributaries: from Mariner 9 (Courtesy NASA)

Plate 14 The Bonn Radio Telescope (Courtesy Max Planck Institut für Radioastronomie)

Plate 15 The Arecibo Observatory, which is part of the National Astronomy and Ionosphere Center operated by Cornell University under contract with the National Science Foundation

the so-called carbon–nitrogen cycle during its residence within the stars. Material that has *not* been processed by these nuclear reactions has an excess of oxygen over both carbon and nitrogen, and it is in material of this type that polysaccharides would form. For material processed by the carbon–nitrogen cycle, on the other hand, there is an excess of nitrogen over carbon, with little oxygen, and in the outflows of such processed material nitrogen-bearing rings which are required for nucleic acids, and also for the C_4N rings in chlorophyll, would arise.

The unidentified dielectric component of interstellar dust referred to in the previous chapter may now be identified with polysaccharides. Provided the polysaccharide has dimensions of the order of the wavelength of visible light, most of the required properties of the dust are satisfied, although in two particular respects we must still turn to material with nitrogen rings. The hump at 2200 Ångströms in the ultraviolet absorption of starlight, which we discussed in the previous

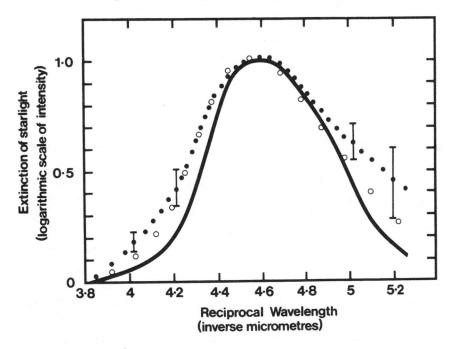

Fig. 20 The 2200 Ångström (4.6 inverse micrometre) band due perhaps to nitrogenic ring compounds with the formula $C_8H_8N_2$ compared with astronomical data on the absorption of starlight

chapter, can be explained by a substance with a C_4N_2 ring bonded to a subsidiary carbon ring, giving an empirical chemical formula (with H atom attachments) of $C_8H_8N_2$. Fig. 20 compares the observed interstellar hump at 2200 Ångströms with the average opacity of the different arrangements of $C_8H_8N_2$. The ultraviolet observations are explained by a mixture of $C_8H_8N_2$ together with some graphite particles, which are still required to account for the overall breadth of the observed hump.

The second requirement for nitrogen-bearing rings comes from the observed spectra of highly reddened stars – in other words, stars whose light penetrates a great deal of dust on its journey to the Earth. Such stars have some thirty diffuse absorption features at well-defined wavelengths between 4400 and 7000 Ångströms. The widths of these features, ranging from 2 to 30 Ångströms, are much too large for them to be attributed to atoms, ions or simple molecules. The most prominent of these bands occurs at 4430 Ångströms and has a width of about 30 Ångströms. Although these features have been studied for more than four decades, their origin has remained obscure. A few years ago, however, F. M. Johnson suggested that porphyrins, like the molecule with the empirical formula $MgC_{46}H_{30}N_6$, could explain many of these mysterious features. Porphyrins, it will be recalled, form the basic structure of chlorophyll, one of the most essential of all biochemical substances. Although Johnson's suggestion has hitherto been considered implausible, in the light of the present discussion it hardly seems so. The union of four C_4N rings to form the central core of a porphyrin molecule is energetically profitable, and for this reason we may expect a fraction of material in nitrogen-rich mass flows to condense into such molecules.

In this chapter we have argued that there is evidence for an astrophysical origin of several basic biochemical units. These include polysaccharides (evidence provided by the 3, 10 and 18 micron bands in galactic infrared sources), nitrogen-bearing rings which could form the purine and pyrimidene bases in nucleic acids (interstellar feature at 2200 Ångströms) and nitrogen-bearing rings assembled into porphyrins (4430 Ångström band and other diffuse interstellar features). With the formation of these materials the foundations of biochemistry would appear to have been laid.

Comets: visitors from distant space | 11

Even if we now know about the existence of the basic chemicals of life in interstellar space, the question arises of how they reached our planet – and perhaps others. What has brought this material from comparatively great distances? In searching for answers we shall look at comets and see how their enormous orbits endow them with the ability to transport material from distant far-flung regions to the inner part of the solar system, and so to the Earth.

Comets are undoubtedly the most spectacular and awe-inspiring of all astronomical phenomena. Reports of cometary apparitions date back to very early times, and recorded observations of them by Chinese and Babylonian astronomers stretch back over four millennia. Numerous references to comets are also to be found in classical Greek and Roman literature as well as in the more recent literature of Western Europe: An example from Shakespeare is to be found in *Julius Caesar*:

> 'When beggars die there are no comets seen;
> The heavens themselves blaze forth the death of princes.'
>
> (Act ii, Sc. 2)

Here, as indeed among all peoples generally, comets are regarded as omens of disaster.

Although observations of comets appear to have been made over a long period of our history, it was only relatively recently – in the last four centuries – that they were recognized as objects outside the Earth. Earlier they were widely regarded as phenomena associated with our local terrestrial atmosphere. Tycho Brahé was perhaps the first to recognize that comets were not connected in any way with the Earth's atmosphere, while the 'periodicity' or regularity of cometary appearances was first noticed by the English astronomer E. Halley (1656–1742). He found close similarities between the orbits of three comets which appeared successively in 1531, 1607 and 1682, with intervals between appearances of approximately seventy-six years. He concluded that these three appearances were of one and the same object which was moving in an almost elliptical orbit around the Sun. By using Newton's recently propounded theory of gravitation and allowing for the effects of perturbations caused by other planets, the French astronomer A.-C. Clairaut (1713–1765) predicted that the same comet would make its next appearance in 1759. His prediction was fulfilled. The comet, now called Halley's comet, is an example of a 'periodic' comet, one which has an average orbital period of about seventy-six years. On the basis of various historical sources, earlier appearances of Halley's comet can be traced back to the first century BC. The most recent appearance was in 1910, and it will appear again in February 1986. The orbit of Halley's comet in relation to planetary orbits is shown in Fig. 21.

While the orbits of planets around the Sun are all nearly circular and confined to one plane, cometary orbits are elongated and inclined at all angles to the plane of the solar system, with the long axes of the orbits in more or less random directions. The elliptical shapes of the orbits vary a great deal from one comet to another. Some comets – the long-period comets – have very eccentric orbits that take them in their courses way beyond the distances of the outermost planets of the solar system, and the orbital period for such comets can be very long, as much as hundreds of thousands of years or even longer. After a number of transits through the inner regions of the solar system a long-period comet becomes eventually transformed into a much shorter-period comet because of the build-up of gravitational interactions with the major planets. A comet with a period of four years would go out again nearly as far as the orbit of Jupiter, whereas one

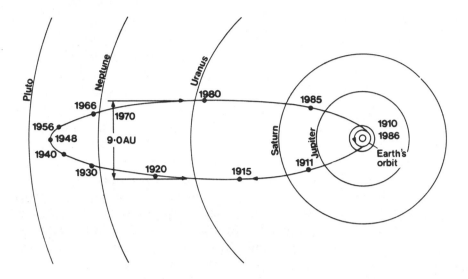

Fig. 21 Orbit of Halley's comet in relation to planetary orbits

with a period of forty years goes out much further, to the orbits of
Uranus and Neptune.

Two or three new long-period comets are discovered every year,
often by amateur astronomers who keep a careful vigil of the night
sky. After a comet makes its first appearance it usually remains visible
for many weeks, streaming slowly against the background of distant
stars as it pursues its orbit around the Sun. A comet becomes visible
only when it comes sufficiently close to the Sun, where it remains for
only a small fraction of the orbital period. The source of this transient
visibility and grandeur is the cometary nucleus, a lump of rather
volatile material measuring no more than ten kilometres across. The
nucleus sheds matter in the form of gas and particles, and when this
matter fluoresces and reflects sunlight it produces a visual display.

As a cometary nucleus approaches the Sun and reaches a point about
three times the radius of the Earth's orbit from it, a diffuse glowing
nebulous envelope or 'coma' usually develops and grows rapidly in
size. This is caused by the evaporation of material from the nucleus. In
a typical case the radius of a fully developed coma could extend to
several hundred thousand kilometres.

Some of the material which evaporates from the nucleus streams

101

away from it to form a cometary tail. The tail keeps its shape because of the pressure exerted upon the evaporating gaseous material by the solar wind, which is an outward flow of particles emanating from the Sun. Cometary tails occasionally make dramatic visual displays extending across much of the sky. The physical length of a comet tail can extend well over a million kilometres, as it did for the most recent appearance of Halley's comet, but most cometary tails are visually not so spectacular, and for any comet the visual extent of a tail is one of the least predictable properties.

The recent long-period comet Kohoutek which appeared in 1973 was a disappointment because the prediction of a spectacular visual display was not fulfilled. From a technical point of view, however, Kohoutek was explored more thoroughly than most earlier comets, and this led to many important new discoveries. Plate 10 is a photograph of Kohoutek showing a tail with much fine structure, probably associated with an interaction between ionized gas from the comet and an interplanetary magnetic field associated with a wind of particles from the Sun.

In recent years comets have been actively studied with the help of many of the latest observational techniques available to modern astronomy. Comet Kohoutek, for instance, was examined at wavelengths in the ultraviolet, visual, infrared and radio wavebands. A number of emission bands corresponding to many atomic species and to simple molecules such as carbon (C_2), cyanogen (CN), methylidyne (CH), the hydroxyl radical (OH) and the amino group (NH_2) had already been discovered in comets by previous simpler studies in the visual waveband. Cometary tails had been shown to have the characteristics of sunlight scattered from small solid particles about a micron in size. These solid particles are thought to be mixed in with the more volatile material which makes up the bulk of the nucleus. As the volatile material is expelled from the nucleus, it carries the dust along with it. The gas and dust in the tail then often bifurcate to give a tail of gas that is separated from the dust tail.

In the ultraviolet spectral region the recent detection of the Lyman alpha line of hydrogen enabled astronomers to study the development of an expanding hydrogen coma. At infrared wavelengths, thermal radiation from a heated 'dust' component was detected. The presence of at least two organic molecules, methyl cyanide (CH_3CN) and hydrogen cyanide (HCN), were also inferred from radioastronomical observations of Comet Kohoutek. (There were several other radio-frequency lines presumably arising from organic molecules which have not yet been identified.) In sum, then, there is by now an impressive list of atoms, ions and molecules which have been detected in comets (see Table 11.1).

Table 11.1 Atomic and molecular species in comets
(See Appendix 1 for names of chemical symbols)

Atoms:	H, C, Si, Fe, Mn, Ca, Na, Cr, Ni, Cu, K, Co, Y.
Radicals, ions and simple molecules:	CN, C_2, C_3, CH, OH, NH_2, CO^+, CH^+, CO_2^+, N_2^+, OH^+, H_2O^+, Ca^+
Organic molecules:	CH_3CN (methyl cyanide) HCN (hydrogen cyanide)
'Dust':	micron-sized metal or graphite grains, rock dust (magnesium silicates), polysaccharides and related organic polymers

The most elusive component of a comet is its nucleus. This is not directly visible even with a telescope, so that the existence of the nucleus has had to be inferred from the atoms and molecules actually observed in the coma and tail. The molecules, radicals and ions observed in the coma are unlikely to be stored in these forms for any length of time because their chemical structures are inherently unstable. It is much more likely that small radicals and ions are fragments of larger molecules which break up as they evaporate and become exposed to ultraviolet radiation from the Sun. The size of the coma depends largely on particular molecules, and especially on the rate at which they flow out of the nucleus, and on their lifetimes in the face of break-up by the Sun's ultraviolet radiation. It is therefore possible that many complex organic molecules exist in gaseous form so close in to the nucleus that they are difficult to detect.

Some years ago, when small radicals were all that had been found in comets, it seemed reasonable to suppose that a mixture of water (H_2O), ammonia (NH_3) and methane (CH_4) served as the source of 'parent molecules', and this led to F. L. Whipple's icy conglomerate model for the nucleus. The almost simultaneous appearance of all the coma radicals, of varying volatilities, at a distance of about three Earth orbit radii from the Sun seemed to demand a somewhat specialized lattice structure for the nucleus. A single type of host lattice molecule appeared to be needed, so that the disruption of the lattice could lead to the release of a wide range of different types of radical. A water-ice lattice was proposed, together with the somewhat arbitrary assumption of an enhanced binding for 'guest molecules' such as ammonia and methane, which were reckoned to be trapped in deep potential well cavities within the nucleus. This so-called 'clathrate hydrate' was also given an arbitrary evaporation temperature considerably above that for water-ice under interplanetary conditions. Furthermore, the evaporation of the 'clathrate hydrate' was believed to release all the ions in the coma as well as any solid particles that happened to be contained in the lattice.

The assumption of a 'clathrate hydrate' lattice appears to us arbitrary and unsatisfactory, particularly in view of the discovery of more complex molecules in the coma. A better explanation in our view is that many of the radicals which are observed are dissociation products of organic polymers, such as the polysaccharides. A mixture of ices, organic polymers, silicate and graphite grains is held together in the cometary nucleus, which becomes eroded by the action of solar radiation when it approaches close enough to the Sun. Hydrated polysaccharides could provide a single 'parent molecular species' for many of the observed radicals and molecules in the coma. This includes ionized water (H_2O^+), the free radicals OH, NH_2, CH, as well as molecules such as diatomic and triatomic carbon (C_2, C_3), the last two being the result of the break-up of pyran rings.

Comets lose material in the form of evaporating gas and dust at every passage past the Sun. They have their orbits continually perturbed by the planets and are therefore forced into ever shorter periods and more frequent journeys past the Sun. In this way comets are stripped of their matter at an increasing rate, so that new comets must replace those that become exhausted. Comets which appeared as long-period comets at earlier geological epochs – for example, at the time of Fig Tree Chert – must have long since dispersed.

How do comets originate? There is no consensus on this matter, but a theory which has gained popularity is one put forward by the Dutch astronomer J. H. Oort. He suggested that a reservoir of cometary material in the form of a 'cometary cloud' envelops the entire solar system at a radius of about one light year, a good fraction of the distance to the nearest star. According to the theory, the cloud contains more than a hundred billion comets. Perturbations caused by the gravitational effects of passing stars cause a small fraction of this vast ensemble to plummet towards the inner regions of the solar system. On this picture, comets must be regarded almost as genuine interstellar voyagers bringing in material from relatively distant regions, since members of the cloud would continually be picking up new material from interstellar space as they cruise through the galaxy along with our solar system. Similarities in composition between cometary and interstellar matter are therefore to be expected, and at least occasionally we may be recipients of substantial amounts of interstellar bounty.

The mass of a cometary nucleus cannot be more than about one billionth of the mass of the Earth, but if a nucleus were to collide with the Earth, the effect would not be negligible. In the present era such an event would cause vast devastation of a metropolis if it chanced to hit one. From modern cometary data we can infer that the probability of such an event is very small, though it cannot be totally ignored. The probability is comparable to that which has been calculated for the catastrophic escape of radioactive material from a fast-breeder reactor. Whereas the latter would kill some tens of thousands of people, a cometary nucleus hitting a major city would kill several millions. People with ample time for worry on their hands should trouble themselves much more about comets than about disasters to fast-breeders. The 1908 meteoritic impact in Siberia, to which we have already alluded, may have been a cometary nucleus. In much earlier geological epochs we believe that the Earth would have been hit far more frequently. Indeed, the Earth could have acquired all of its volatiles – including all the oceans – from such collisions. And, of course, the presence of organic prebiotic chemicals such as we have discussed would have led to a vast input of life-forming materials to the Earth.

Meteorite clues | 12

Some of the most important clues about the origin of life in the universe come from meteorites, which could have been related to comets in the early history of the solar system. One of the earliest records of a possible meteorite shower is to be found in the Bible (Joshua 10 : 11), where there is a description of a fall of 'great stones from heaven' upon the fleeing Amorite tribesmen. Solid bodies which enter the Earth's atmosphere from interplanetary space are collectively called 'meteoroids'. Speeds of incoming meteoroids vary from about thirteen to seventy-two kilometres per second, depending on the relative motion of the Earth and meteoroid, the fastest speed corresponding to a head-on collision. If they are large enough, meteoroids can give rise to 'meteors' or shooting stars, a phenomenon caused by frictional heating brought about by collisions with atmospheric molecules.

Meteoroids start incandescing to show visible trails (Plate 11) at heights of between seventy-five and fifty miles above the Earth's surface, the faster meteoroids becoming luminous at the higher altitudes. A meteoroid more massive than about ten kilograms survives its journey through the atmosphere to land on the Earth as a meteorite. Only a relatively thin surface layer is removed and the interior of the object is not significantly heated in travel. Still larger meteoroids often break apart in mid-air and produce showers of stones. The smallest micron-sized meteoroids do not become seriously heated in transit, however, and drift down through the terrestrial atmosphere to settle eventually at the Earth's surface. Meteoroids of intermediate size (greater than 0.1 mm) burn out completely while they are still high in the atmosphere and so merely produce the well-known 'shooting star' appearance. They are seen in the night sky with an average frequency of six to ten every hour. At certain well-defined times in the year this rate is increased more than tenfold for several days on end, when the Earth is passing through accumulated trails of particle debris left by comets after numerous passages near the Sun. The total mass of meteoritic matter which settles on the Earth every day is estimated at about a hundred metric tons, the main contribution coming from micrometeorites.

The much larger meteorites beat down upon us at a fairly steady rate. Over two thousand of them are estimated to fall on the Earth every year, but only very few – less than a dozen – are tracked down and the samples retrieved. The vast majority of falls occur on oceans or in remote areas and go undetected. The average weight of a meteorite find is several kilograms, but very large falls do sometimes occur, as in the case of the Allende meteorite which fell in 1969 and led to a find of over a thousand kilograms. Meteorites from over 1500 falls have been collected, and the total mass of this material which is carefully preserved in our museums exceeds several hundred tons.

On the basis of chemical and petrological studies, meteorites are now known to come in three main forms, depending on whether they are made entirely of stone, iron, or a mixture of the two. The stone meteorites account for over ninety per cent of all finds and we shall be concerned with a particular type of them, the carbonaceous chondrites, which make up about three per cent of total finds. The carbonaceous chondrites are of special interest to us because they have similarities to cometary as well as to interstellar matter and they are probably among the oldest objects in the solar system.

Among carbonaceous chondrites three types are characterized by an appreciable content of carbon in the form of organic molecules. One group (type I) consists of meteorites which have sulphur in the form of sulphates, iron as nickelian magnetite, and silicates in an amorphous hydrated form. They have a fine-grain structure indicating a low temperature assembly of separate small particles and their total carbon content is about 3.5 per cent. Type II carbonaceous chondrites, with a carbon content of 2.5 per cent, contain elementary sulphur, as well as iron in the form of a hydrated silicate. They contain 'chondrules' or insertions of material measuring up to one centimetre across. The other type of carbonaceous chondrite (type III) is generally similar to type II but has much less carbon (about 0.5 per cent), less hydration and a less conspicuous chondrule structure. (Details of a few of the more important falls are set out in Table 12.1.)

Table 12.1 Details relating to some carbonaceous chondrite falls

Meteorite	Type	Site of Fall	Date	Mass of Find
Orgueil	I	France	14 May 1864	10 kg
Ivuna	II	Tanzania	16 Dec. 1938	0.7 kg
Mokoia	II	New Zealand	26 Nov. 1908	45 kg
Murchison	II	Australia	28 Sept. 1969	225 kg
Murray	II	U.S.A.	20 Sept. 1950	7 kg
Nagoya	II	Argentina	3 June 1879	5 kg
Allende	III	Mexico	7 Feb. 1969	> 1000 kg
Vigarano	III	Rumania	22 Jan. 1910	15 kg

Radioactive dating techniques give an age of formation for the carbonaceous chondrites of between 4.5 and 4.7 billion years, so that they are older than the Earth's crust. There can be little doubt that these objects have undergone the most gentle possible thermal history since the time of their compaction from separate grains and molecules. At no time in their history have they been heated to temperatures of more than 400–500 degrees above absolute zero, for any heating above these temperatures would have certainly altered their primordial chemical and mineral structures which persist to this day.

Recent studies of particles extracted from type II carbonaceous chondrites have shown the presence of micron-sized clumps consisting of many separate grains each with a radius of about a hundred Ångströms. There is also evidence that a component of the same meteorites is genuinely extrasolar, having condensed from a gaseous state in distant parts of the galaxy. This conclusion is drawn from the detection of several isotopic ratios which are distinctly anomalous – such as the ratio of one isotope of the gas neon to another ($^{20}Ne/^{22}Ne$) in a component of the meteorite. The observed isotopic anomalies (relative to solar-system values) are explained if grains condensed in the vicinity of a nova or supernova. We showed some years ago that such condensations occur within the gaseous matter which flows as an expanding envelope from out of such an object. These grains, which are embedded in the chondrites as clumps, did not subsequently remelt and they still carry the imprints of certain extinct radio-activities derived at the source of their original condensation. We therefore have a most tangible solid particle component of interstellar space arriving intact within meteorites. If this is granted, we may suppose that the organic matter associated with carbonaceous chondrites is also of interstellar origin.

More than two per cent of the carbon in type I and type II carbonaceous chondrites is in the form of a variety of organic compounds, which may be separated into two groups: those which are soluble in organic or inorganic solvents, and those which are not. The insoluble component has proved the more intractable and controversial. An intriguing controversy followed a report by G. Claus and B. Nagy in which they claimed to have detected structures resembling fossil microbes in carbonaceous chondrites. Organized structures with dimensions ranging from four to thirty microns, vaguely resembling fossil algae, had first been discovered in the Orgueil and Ivuna meteorites, both of which were actually seen to fall, tracked down and recovered, the Orgueil meteorite in France in 1864, the Ivuna meteorite in Central Africa in 1938. Similar fossil-like organized structures were later found in a few more carbonaceous chondrites, but other types of meteorites do not show these features.

Five different types of fossil-like structures have been observed, of which four vaguely resemble certain types of single-celled terrestrial organisms living in water, such as algae. Differences of detail, however, make it difficult to identify the 'fossils' with any known terrestrial species. Electron micrographs of these 'fossils' show that most of them are roughly spherical, and there is evidence of a substructure resembling cell walls, as well as residues of cell nuclei pores and flagella-like formations. The fifth type of fossil was found to be totally unlike any known terrestrial organism. It is roughly hexagonal, with three of the bounding surfaces thicker than the others, and the entire hexagonal structure is surrounded by a spherical halo. In several cases the fossil structures observed by Claus and Nagy show constrictions in the central parts of elongated objects, and this suggests a process similar to cell division.

Strong arguments to support the view that these fossil structures are indigenous to carbonaceous chondrites were put forward by Claus, Nagy and Harold Urey, among others. The concentrations of fossils are remarkably high in all cases studied and account for about ten per cent of the insoluble organic matter in these meteorites. If these structures are indeed extraterrestrial life forms, the implications for the origin of life would be profound. We would then have evidence of organisms which evolved outside the confines of our planet – evidence which would be highly relevant to the argument we have developed. Many questions, however, remain to be answered, the most immediate one being how such organisms could evolve and become fossilized within the fabric of meteoritic rock. In the mid-1960s the lack of a convincing answer to this question led many scientists to become sceptical about the fossil explanation of the so-called 'organized structures' in meteorites.

An alternative explanation was that these fossil-like structures are mineral grains which have acquired coatings of organic molecules by some non-biological process. The difficulty with this theory, though, is that the highly organized cell-like appearance of these structures would still remain a mystery. Terrestrial contaminations were suggested as another possibility, but this is unlikely to be the correct explanation for the majority of structures, because they have no terrestrial fossil counterpart.

Incredible and revolutionary though it might seem, a biological origin for the Claus–Nagy structures appears possible, and the cometary model discussed in Chapter 14 would lend some support to this hypothesis. We shall argue that primitive living organisms evolve in the mixture of organic molecules, ices and silicate smoke which make up a comet's head. The volatile inorganic ices are slowly boiled off in the numerous journeys of the comet past the Sun, leaving a concentrated soup of prebiotic molecules and silicate dust in which living organisms evolve. Type I carbonaceous chondrites are therefore most probably 'spent' comets. The ices and most of the organics too have been vaporized, but compacted silicate grains along with some organic polymers and (possibly) fossil organisms remain.

Recent studies of the bulk of the insoluble organic matter in carbonaceous chondrites have shown that it has a highly condensed coal-like aromatic structure, with an infrared spectrum which is not very different from that of cellulose. We believe that this matter represents a modified form of interstellar cellulose – coalified, just as coal is formed from wood under terrestrial conditions.

In 1963 B. Nagy and C. M. Bitz reported the presence of fatty acids in the Orgueil meteorite, and nitrogen bases including purines and pyrimidines were found in it later. Pigments resembling porphyrins have also been detected in several carbonaceous chondrites, and this implies that carbonaceous chondrites contain many if not all of the nitrogen-bearing ring structures that appear in biology.

Early reports of amino acid detections in meteorites were discredited when it was discovered that human fingerprints give an equally rich amino acid spectrum, but later experimenters used a 'contamination-proof' technique to show that seventeen amino acids were indigenous to the Murchison meteorite. Ten of these were not found in terrestrial proteins. Table 12.2 summarizes recent amino acid detections in three carbonaceous chondrites, giving only those cases which are biologically significant. The average content of amino acids in meteorites is about fifteen parts per million. Although this might seem only a modest amount, it is a much greater fraction than could be

Table 12.2 Amino acid compositions of carbonaceous chondrites (non-biological amino acids excluded)

Amino Acid	Murchison	Murray	Nagoya
	(per cent of total amino acids)		
Aspartic acid	3.4	5.5	10.1
Glutamic acid	6.6	4.8	20.3
Glycine	33.6	17.7	27.6
Alanine	14.0	6.6	7.8
α-Aminoisobutyric acid	19.4	50.7	0
β-Alanine	6	5.7	11.9

expected for the outer crust and atmosphere of the Earth due to 'energizing' events like thunderstorms.

A further link between carbonaceous chondrites and interstellar matter is provided by the interstellar spectral signature at 2200 Ångströms which we have already discussed. Together with A. Sakata, N. Nagakawa, S. Isobe, T. Iguchi and M. Morimoto from Japan, we obtained an ultraviolet transmission spectrum of soluble organic matter extracted from the Murchison meteorite. The spectrum of this extract showed an absorption feature at precisely the same wavelength as the interstellar band at 2200 Ångströms and had the same width. Closely similar chemical species – probably nitrogen rings – are therefore present in carbonaceous chondrites as well as in interstellar space.

It has been argued in recent years that organic molecules could form under primitive solar system conditions by reactions involving carbon monoxide (CO), hydrogen (H_2) and ammonia (NH_3), with surfaces of magnetite, nickel-iron, and hydrated silicates acting as catalysts. This process, known as the Fischer-Tropsch synthesis, was discovered in 1923 and is used industrially for the production of gasoline. It is highly unlikely, however, that such reactions, which operate under high density conditions at pressures of 0.1–10 atmospheres, can operate at all under the presumed early solar system conditions of very much lower density. In view of the strikingly close similarities of chemical composition between interstellar matter, comets and carbonaceous chondrites, we consider it far more likely that the carbonaceous matter in these meteorites represents a primordial interstellar component. In other words, the molecules of life have reached this planet through the addition of cometary and meteoritic materials, which – as we have discussed in detail in earlier chapters – themselves reflected the organic materials still pervading the interstellar clouds of gas and dust.

The birth of the solar system | 13

The Sun and its family of nine planets, their attendant moons and the asteroids were formed from a fragment of an interstellar cloud close to five billion years ago. The processes which led to the formation of the planets are of particular relevance to our story. Besides satisfying a natural curiosity to know how our own particular abode came into being, these processes can shed light on how many Earth-like places we could expect to find elsewhere in the universe.

We know that stars are forming more or less continuously from interstellar clouds. A typical cloud has a diameter of about fifteen light-years and a total mass of several thousand times that of the Sun. Stars are spawned not singly but in groups or associations of many hundreds or even thousands. It is likely that the Sun was one of about a thousand stars which formed at roughly the same time by the collapse and fragmentation of an interstellar cloud. The collapsing cloud fragment that eventually ended up as our solar system was more massive than the Sun, though not by a large amount – in fact by only about ten times the present mass of all the planets, or about one per cent of a solar mass.

How is it that not all the matter in this cloud fell into the Sun? How could planets separate out from such a collapsing fragment? In attempting to answer these questions over the years, astronomers have tended to develop two main types of theory, known as 'catastrophic' and 'quiescent'. In one form of the catastrophic theory the tidal interaction of a passing star with the Sun was thought to rip off the materials which eventually formed planets. In another version the Sun was a member of a binary star system, with the binary companion exploding and leaving behind a wisp of material which went to form the planets. As for quiescent theories, the essential feature of most of them dates back to Laplace (1749–1827). A gaseous planetary disc is said to have developed naturally in a collapsing, rotating prestellar gas cloud, with planets condensing from the disc's gaseous material.

It is now generally accepted that the planets formed in a quiescent rather than a catastrophic process. The Sun as well as the planets condensed from a single cloud fragment, though the precise details are still the subject of controversy. The salient features of the present solar system are summarized in Table 13.1.

Table 13.1

	Distance from Sun (relative to Earth's distance)	Radius (relative to Earth's radius)	Mass (relative to Earth's mass)	Period of revolution, Earth years	Period of rotation or spin, Earth days	Orbital speed (miles per second)	Density (water = 1)
Mercury	0.39	0.4	0.056	0.24	59.0	30.0	5.0
Venus	0.72	1.0	0.8	0.62	243	22.0	5.3
Earth	1.0	1.0	1.0	1.0	1.0	18.0	5.5
Mars	1.52	0.5	0.11	1.88	1.03	15.0	4.0
Jupiter	5.2	11.0	318	12.0	0.41	8.0	1.3
Saturn	9.5	9.5	95.0	29.5	0.43	6.0	0.75
Uranus	19.2	3.5	14.5	84.0	0.45	4.5	1.5
Neptune	30.1	3.5	17.0	164	0.66	3.3	1.7
Pluto	40.0	0.5 (?)	0.1 (?)	247	0.27	3.0	?

In addition to the nine major planets listed in Table 13.1, there are many minor planets or asteroids – on the average no bigger than large boulders – with a total mass of only 0.02 per cent of an Earth mass in a belt of orbits located between Mars and Jupiter. Although the planets Mercury and Venus have no attendant satellites, the Earth has one,

Mars has two, Jupiter has thirteen, Saturn has ten, Uranus five and Neptune two. The orbits of all the planets are practically circular, and they lie almost in one plane, which is roughly at right angles to the axis of the Sun's spin (Fig. 22).

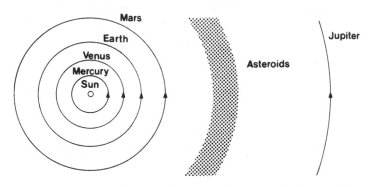

Fig. 22 Planetary orbits in the solar system (schematic diagram)

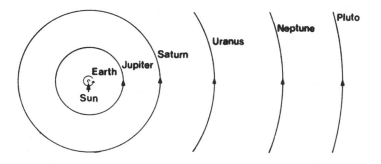

Table 13.2

Group:	Terrestrial planets	Intermediate 'gas giants'	Outer planets
Planets:	Mercury, Venus, Earth, Mars (Asteroids)	Jupiter, Saturn	Uranus, Neptune
Composition:	Mainly comprised of rock and iron; magnesium, silicon, iron	Mainly comprised of hydrogen, helium, probably have terrestrial cores	Mainly comprised of carbon, nitrogen, oxygen as methane, ammonia, carbon dioxide, water

Planets, which can be broadly divided into three main types according to the criteria set out in Table 13.2, raise several important questions which any viable theory of planet formation must be able to answer satisfactorily. As the inner planets are made up of materials such as iron and rocks which have high melting temperatures, while the outermost planets are made up of volatile substances such as water, methane and ammonia which have a comparatively low melting point, how did this chemical segregation occur? How did the Earth acquire its small quantities of volatiles and why are there certain conspicuous anomalies in their relative abundances? Why, among the volatiles, are the rare gases neon, krypton and xenon scarce in comparison with oxygen, carbon and nitrogen?

A third area of difficulty concerns the Sun, which spins remarkably slowly on its axis – once every twenty-six days. All mass elements in the solar system are said to have a physical quantity known as angular momentum, which is mass multiplied by velocity multiplied by distance from an axis of spin. This idea of angular momentum is important for the reason that when all the bits of angular momentum of all the mass elements of the whole solar system are totalled together, the result must be an unchanging quantity. The strange thing is that the Sun carries only two per cent of this unchanging total, and yet the Sun has 99.9 per cent of all the mass. How did this peculiar distribution come about? These are just a few of the questions that need answering in any rational theory of the origin of the solar system.

A variant of a model originally discussed by Laplace seems to offer the best hope of a solution. A gas cloud began to shrink under its own gravity (see Fig. 23). It started off as a nearly spherical blob of gas spinning very slowly on an axis at the rate of about one turn in every ten million years. The total mass of the blob was about 1.01 times that of the present Sun and its radius was about one tenth of a light year. This means that the gas had a density of about half a million hydrogen molecules per cubic centimetre, a figure typical of the dense cores of many molecular clouds in regions of star formation throughout the galaxy.

As the cloud shrank in the absence of external forces, it was able to spin more and more rapidly and became flattened at its poles like a squashed orange. The laws of dynamics require that angular momentum is preserved during the shrinkage, and this in turn implies that the

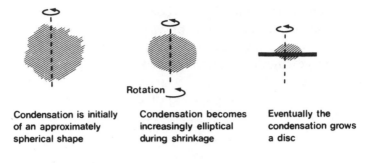

Condensation is initially of an approximately spherical shape

Rotation

Condensation becomes increasingly elliptical during shrinkage

Eventually the condensation grows a disc

Fig. 23 Birth of the solar system: rotating gas sphere flattening and developing a disc as the speed of rotation increases

speed of a gas particle at any point on the equator varies in inverse proportion to the equatorial radius. At the equator the rotation speed of a cloud fragment at the start would be close to two metres per second. Collapse of such a fragment to the size of the present Sun would mean shrinkage by a factor of a million, and so imply a millionfold increase of the equatorial speed. By this argument the Sun ought to be turning on its axis with an equatorial speed of about two thousand kilometres per second. As it happens, however, the present-day Sun has an equatorial speed of only about two kilometres per second and is therefore rotating a thousand times too slowly to account for its initial complement of angular momentum.

The most probable solution to this mystery is that the missing angular momentum was 'shed' along with the small fraction of matter which went into forming the planets. Indeed, some such shedding was essential, since the present Sun simply could not rotate as fast as two thousand kilometres per second at the equator without being violently torn apart by rotary forces, in much the same way that a flywheel bursts apart if it is spun too rapidly. The contracting blob of gas which went to form the solar system would have suffered such a fate when its radius was about two-fifths of the orbital radius of the innermost planet, Mercury. At this stage of contraction the so-called centrifugal force on an element of gas at the Sun's equator exceeded gravity, so resulting in the ejection of a disc of material from the equator. The expelled disc remained connected to the central solar condensation by a magnetic field which acted as a torque transmitter, so that one should picture something rather like spokes connected to the centre of a wheel (Fig. 24). The magnetic field may be assumed to have been torn apart from the Sun along with ionized gas, with the result that the magnetic field transferred angular momentum from the Sun to the disc as the

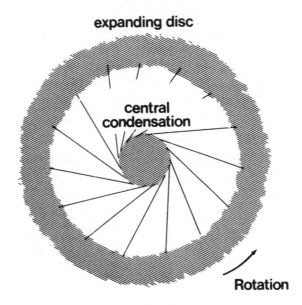

Fig. 24 Birth of the solar system: magnetic 'spokes' transfer angular momentum from central condensation to the expanding disc

latter expanded outwards to the orbital distances of the outer planets. The strength of the magnetic field necessary to bring about this transfer in the available timescale was about one twentieth of the field strengths observed in present–day sunspots. So this estimate would not be unreasonable.

If the Sun has lost a large part of its angular momentum, what about the angular momentum of the solar system's planetary material? To resolve this question correctly we must adopt an appropriate definition of 'planetary material', taking account of gas in the disc which did not become incorporated into present–day planets. The inner terrestrial planets (Mercury, Venus, Earth and Mars), as well as the outermost planets (Uranus and Neptune), are notably lacking in the cosmically most abundant elements, hydrogen and helium. In

Table 13.3 we list the masses of the three major groups of planets, as well as their main chemical constituents, and we give multiplicative factors for augmenting these masses to recover the original masses of gaseous material which were associated with these planets.

Table 13.3

	Jupiter *Saturn*	*Uranus* *Neptune*	*Terrestrial* *Planets*
Main constituents and their percentages relative to solar composition	hydrogen, helium (100%)	carbon, nitrogen, oxygen (1.5%)	magnesium, silicon, iron (0.5%)
Present masses ($M\odot/10^5$)	120	10	1
Multiplier	$\sim$1	67	200
Augmented masses ($M\odot/10^5$)	120	670	200

The bottom line of Table 13.3 gives the masses of planetary material when their present composition has been augmented with enough hydrogen and helium to make up a full solar complement of these gases. It is known that the elements carbon, nitrogen and oxygen make up about 1.5 per cent of the mass of the Sun. So for Uranus and Neptune, whose masses come mainly from molecules built from atoms of carbon, nitrogen and oxygen, the present masses are to be multiplied by the factor 67. The inner planets are mainly comprised of the elements magnesium, silicon and iron, which make up about 0.5 per cent of the mass of the Sun, and the corresponding augmentation factor is 200. Jupiter and Saturn are the only two planets which have approximately their full complement of hydrogen and helium. We find that whereas the present total mass of the planets is about a thousandth of a solar mass, the original mass of the material associated with the planets was close to a hundredth of a solar mass. This higher mass of planetary material must have been squirted out from the fast-spinning solar condensation when gravitational forces were unable to balance the rotary forces in the equatorial regions.

When such a rotary crisis occurred, the radius of the Sun was about thirty times its present size and its surface temperature was close to 3500 degrees. Nuclear reactions had not yet begun, and a high initial luminosity, amounting to 150 times the present solar luminosity, was derived from the dissipation of the energy from gravitational collapse. The disc of material ejected from the equatorial regions was magnetically coupled to the parent condensation but expanded outwards from the Sun, cooling as it did so. Droplets of liquids and fine smoke particles condensed from the cooling gas. The first major condensates to form when the gas cooled to about 1500 degrees, the sort of temperature found in a blast furnace, were droplets of iron and particles of mineral silicates. This temperature was reached when the expanding gaseous disc crossed the present orbital distances of the inner terrestrial planets. A fraction of the iron droplets and silicate smoke particles coagulated into metre-sized solid objects which 'fell out' of the expanding gas and settled into nearly circular orbits at the distances of the terrestrial planets.

A residual fine-smoke component which did not coagulate into large enough particles was nevertheless carried outwards along with the hydrogen and helium and other volatile gases by frictional drag. Particles aggregating together at the distance of the asteroidal belt would have fallen out there as meteoroid-sized objects in circular orbits, so making the asteroids. Then the next major condensates to form would have been frozen ices of water, carbon dioxide and ammonia. This further condensation took place at a temperature of about 150 to 200 degrees, a temperature reached at the orbital distances of the present outer planets, Uranus and Neptune. The ices condensed as mantles on small silicate or iron particles which came as smoke particles from the inner part of the disc. These icy particles also quickly coagulated into a swarm of metre-sized blobs which settled into circular orbits near the present distances of the outer planets. The bulk of the hydrogen and helium expelled from the Sun was then lost by evaporation at about the orbits of the outermost planets, the excess angular momentum being also lost along with this evaporation of gas.

We have argued so far that a chemical segregation into metre-sized rocks and icy blobs occurred naturally in the planetary disc, the former at the distances of terrestrial planets and the latter at the distances of the outer planets. A further clumping of the rocks eventually led to the formation of Mercury, Venus, Earth and Mars, while a clumping of ice blobs eventually led to the formation of Uranus and Neptune.

The clumping of metre-sized objects to form planets took place in two stages. In the first stage they were brought together into swarms of cometary objects from ten to a hundred kilometres across, a process which took a mere few thousand years for the inner regions and a few hundreds of thousands of years for the outer regions. Further aggregation leading to planets then occurred when the many small bodies collided as a result of their random criss-cross orbits. This further aggregation took much longer to complete: the four inner planets were formed in about two million years, while the outer planets took a much longer period of about three hundred million years.

Before completion of the final stage of aggregation of the outermost planets, two intermediate-sized objects probably dipped into the orbital distances of Saturn and Jupiter as a result of random deflections and were trapped in stable orbits at these distances. These objects later mopped up hydrogen and helium from the outflowing gas in the planetary disc to become the giant planets Saturn and Jupiter.

The last stage of aggregation of the inner planets involved collisions which were energetic enough to cause impact melting and a chemical segregation of iron from rocks. The denser molten iron sank to the centre of the accumulating planet, and the lighter silicate floated on the surface and quickly re-solidified after each impact. The molten iron core of the present Earth is a relic of this impact-heating phase in its early history.

Two million years after the gaseous planetary disc was expelled from the fast-spinning Sun the formation of the inner planets (Mercury, Venus, Earth, Mars) was nearly complete. Except possibly for Mars, each of these planets had an iron core surrounded by a rocky mantle. Their surfaces were heavily pock-marked with craters produced by impacts which occurred during the last stages of accumulation. All these planets then had almost no volatile materials such as water, carbon dioxide, ammonia, methane and the rare gases.

By now the Sun had cooled from its initial high-luminosity phase. It had contracted to its present radius, started the nuclear burning of hydrogen in its interior and had a luminosity lower than at the present day. The surface of the primitive Earth was dry, heavily scarred and colder than now, and the overall scene would have looked most unpromising for the emergence of life.

Even then, however, the drama of planet formation was not over. The scene shifts to the outer regions of the solar system, where the outer planets had scarcely begun to aggregate. This region was still occupied by a swarm of cometary-type objects in nearly circular orbits. These hundred-kilometre-sized icy bodies which condensed largely from gases in the planetary disc must surely have intermingled to a significant extent with grains and complex molecules that were indigenous to the molecular cloud from which the solar nebula fragmented. The entire solar nebula would have had a random oscillatory motion relative to the 'primitive' gas, and a mopping up of presolar dust and molecules would have occurred on the surfaces of the swarm of accumulating 'comets' out at the orbital distances of Uranus and Neptune, a process that would have been far more effective than the addition of such dust and molecules to the Earth and the other inner planets.

Random deflections of these cometary-type objects would have sent one or more of them occasionally plunging towards the inner regions of the solar system. Several thousand direct encounters of such bodies with the Earth could have occurred during the accumulation period of the outer planets. While in many cases the impacting material would have splashed out again into space, in other cases icy material was retained, to burst apart upon impact and splash itself over a large area. The splashed material would have become partially refrozen, but its more volatile components would have maintained enough of an atmosphere of gas to provide much softer landings for further impacts of smaller bodies. This cushioning or frictional drag brought about by atmospheric gases has allowed meteorites to make soft landings ever since.

What can we now say about the unusually low abundances of neon, krypton and xenon on the Earth? In the case of neon in particular, its mass in the Earth's atmosphere in comparison with nitrogen in the atmosphere and oxygen in the form of water in the oceans is very minimal indeed: one part neon to sixty thousand parts nitrogen to fifteen million parts oxygen. The corresponding ratios for the original solar nebula must have been about one part neon to three-quarters part nitrogen to five parts oxygen. The enormous discrepancies in these ratios cannot be explained by any differences in the process of evaporation, since the masses of nitrogen, oxygen and neon atoms are all fairly similar. The observed low abundances of neon, krypton and xenon must therefore be attributed to outgassing of an original volatile content in rocks. The amount of these gases which could have been trapped during the high temperature condensation phase of the Earth's core and crust was very low, as was the trapping of all volatile materials. The very much larger quantities of carbon, nitrogen and water arrived in the form of icy cometary nuclei from the outer reaches of the solar system in the manner described above. There was no corresponding addition of neon, krypton and xenon because these elements did not condense as ices, even at the low temperatures in the region of Uranus and Neptune.

The picture of the origin of the solar system described here overcomes many of the difficulties involved in alternative explanations which are currently in vogue. For instance, if the Earth condensed from a cooling gaseous protoplanet, it is hard to explain the observed proportions of the volatiles. The amount of water, neon and carbon dioxide would exceed the amount of iron and rock. Yet the mass of the oceans is only 1.42 million billion billion grams, a mere 0.024 per cent of the Earth's mass (5.977 billion billion billion grams), while neon is almost non-existent, as we have just seen. There are so many serious objections to the gaseous protoplanet theory that it cannot be considered seriously. For the origin of life on our planet, therefore, all that was needed was a primitive atmosphere which allowed the soft landing of small cometary bodies carrying interstellar prebiotic molecules. We know that such soft landings of meteorites occur today. In the beginning the solar system would have picked up considerable quantities of cometary-type debris from its parent molecular cloud, as it carried out an oscillating movement within the cloud. The 'beginning' may have continued for tens, if not for some hundreds, of millions of years.

First days on Earth | 14

Our geological history starts with the formation of the Earth's crust about 4.5 billion years ago. The Sun had by then shrunk to approximately its present size and had begun its long life as a main sequence star. It had just started converting hydrogen to helium in its interior and at this early stage of its life the luminous output was some twenty-five per cent less than it is now. As a consequence of the lower luminosity, the inner planets were colder than they are at present. The Earth was also dry and heavily cratered from the impacts which it had suffered during the final stages of its aggregation. In many respects our planet would have looked similar to the present-day Moon. There would already have been a segregation of minerals into strata of varying densities, the lighter rocks floating upon the heavier ones. The lightest overlying rocks would have formed continental rafts, and the slow processes of mountain range building would have been under way.

How did such an arid, dreary, lunar-like landscape become transformed into a habitat for life? How did the Earth acquire an atmosphere of volatile gases, in particular water and carbon dioxide and possibly ammonia? The usual answer is that these volatiles were produced by the outgassing of rocks in the Earth's crust, when volatiles that were initially trapped at high temperatures in rocks slowly leaked out. This explanation, however, is open to serious objections. To begin with, it is not based on any evidence. One might also say it was another attempt at reviving the old idea of 'spontaneous generation' – drawing water from rocks in this case.

If the Earth really was formed from a cold protoplanetary gas and dust cloud, as many people think, it would have kept roughly the original proportions of chemical elements found elsewhere in the universe, such as we now see in the Sun. From such a system we would expect to obtain a great deal more water than rock, whereas we have a mass of only one part water in three thousand parts rock on the present Earth. Where has all the water gone? One might argue that most of it was lost, evaporated away, but why was it not all lost, and what happened to the other volatiles such as the rare gases? Why were neon, krypton and xenon much more depleted than other volatiles? There are so many unanswered questions in this picture. An explanation which is far more rational is that all the volatiles except neon, krypton and xenon, which are chemically inert, came later in our geological history, and from outside, and along with them came also the complex biochemicals required for the origin of primitive life forms. Only the present exceedingly small amounts of neon, krypton and xenon resulted from outgassing from the interior of the Earth.

At the time when the formation of the Earth's crust was complete, the aggregation of the outer planets had barely started. The region between the present orbits of Uranus and Neptune was occupied by a multitude of objects about a hundred kilometres in diameter orbiting around the Sun. These objects were made up of a mixture of ices which condensed in the outflowing planetary gas, together with large quantities of interstellar matter – dust and prebiotic molecules – that became mopped up from the dense interstellar material in which the new solar system was embedded.

If interstellar space is full of prebiotic molecules – the type or subset which consists of those molecular structures intimately associated with terrestrial life – it is almost self-evident that the origin of life on Earth merely involved a piecing together of interstellar prebiotics. It is easy to understand why all life turns on basic molecular combinations of hydrogen, carbon, oxygen, nitrogen, sulphur and phosphorus from the billions of possible combinations that might have been equally workable. It is also easy to understand why the same molecular configurations are deployed for a multitude of different functions and do not always make the best use of available energy resources. A case in point is chlorophyll. The sunlight-chlorophyll connection drives the whole of biology, but it comes as a surprise to find that chlorophyll does not have its absorption bands where the Sun emits most of its energy (Fig. 25) – in other words chlorophyll does not make the best use of sunlight. If life had a free choice to pick the best pigment-battery system, it would quite likely find a substance better related to the Sun, our particular star, for this most crucial function. That it did not do so is significant. Life simply used what was available to it from the interstellar warehouse.

Some people argue that all interstellar organics were destroyed in the processes that led to the formation of the Earth, but we know that

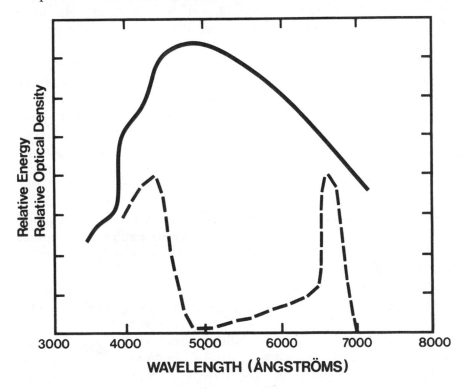

Fig. 25 Spectrum of chlorophyll A (broken curve) compared with energy distribution of sunlight at Earth's surface (solid curve). The energy and optical density scales are arbitrary

comets do carry organic compounds and some meteorites bring such molecules to Earth entirely preserved. Recent investigations have shown that chondrites have optical properties which are inconsistent with an origin in the asteroidal belt. Chondrites resemble more closely objects such as Icarus which lie in orbits crossing that of Mars, and they are therefore more likely to be of cometary origin.

Primitive life forms may even have evolved extraterrestrially on comets. Heat released at some depth below the surface of a comet could have melted a fraction of the underlying ice, the heat being released by chemical reactions between the organic molecules. Moreover, once some ice was melted, there would have been further heat-releasing chemical reactions between the organic molecules and liquid water. Such a situation, adequately insulated against heat loss by overlying surface layers, could well have provided the most favourable conditions for the emergence of life – the best location of all for a primeval soup.

Prebiotic molecules including polysaccharides, amino acids, porphyrins (which are essential to the structure of the chlorophyll molecule) and the component substructures of nucleic acids could all have been present in concentrations within the melted ice. Indeed, the concentrations of these substances would have been far higher in such a mixture than in any terrestrial location. The mixture would already have included the basic life molecules – polynucleotides, polypeptides, porphyrins, carotenoids. According to our point of view, all the essential building-blocks of life would have been there, without any need for further energizing events. We suspect that the first living organisms (prokaryotes), those that could live anaerobically – without free oxygen – were put together in such cometary sites from preexisting macromolecules.

If life originated in the way we have described here, its appearance and proliferation on Earth must have happened relatively quickly. There is evidence for organisms like bacteria and blue-green algae inhabiting our planet about 3.1 billion years ago, less than a billion years later than the oldest known rocks. And there is no reason to believe that the oldest organisms have been found, since they are almost certainly the most difficult to discover.

Recent studies of radioactivity in moon rocks have shown convincingly that catastrophic events occurred on the lunar surface about 3.9

billion years ago, probably involving violent meteoritic impacts. It is difficult to imagine that similar events did not occur at the same time on the Earth's surface, and it may well be that our planet was unfit for the widespread dispersal of life much before 3.9–4 billion years ago. We may then tentatively set the emergence of terrestrial life at four billion years ago – an event heralded by the arrival of a life-bearing comet.

The Earth's atmosphere, however, may have evolved significantly over the preceding half billion years. After the first hard landing of a comet which we discussed earlier, subsequent soft cometary landings would have caused the spreading of volatiles over the Earth's surface. The water would at first have been frozen, but cumulative additions of carbon dioxide could have rapidly changed the terrestrial climate. The total amount of carbon dioxide now present as carbonates in sedimentary rocks must have initially come from comets. There could well have been a carbon dioxide pressure of many atmospheres over the primitive Earth, a circumstance that would quickly have changed our abode from being cold and dry to becoming hot and humid. This is because carbon dioxide in the atmosphere produces a 'greenhouse' effect. Visible light from the Sun is able to pass through the atmosphere and to heat the Earth, but infrared radiation from the heated Earth is prevented from escaping because atmospheric carbon dioxide is able to absorb it. The effect is to lift very considerably the temperature at ground level. As the temperature rises, water-ice melts, water vapour evaporates into the atmosphere, and the water vapour pressure increases steeply with the temperature. Convection sets the atmosphere into motion with rapidly rising and falling columns, enabling water vapour to condense into droplets high up in the atmosphere. The effect of this condensation is to produce a cooling from the upward transfer of latent heat. Such a cooling process, however, would not have become critically important until the oceans had become very warm, perhaps with a temperature above fifty degrees centigrade. This would have favoured the more thermophilic of the cometary organisms, namely those which originated in water that was warmed considerably by the release of chemical energy.

The eventual cooling of the Earth in such a picture would have depended critically on the rate at which atmospheric carbon dioxide became converted into limestone in rocks. Carbon dioxide can be

131

'fixed' into rocks by dissolving in water and by the water then coming in contact with suitable solid material. At the present time these processes are greatly helped by living organisms which can concentrate carbon dioxide in the soil. The carbon dioxide is taken up by rain water, which in turn washes over rocks to produce deposits of limestone. These processes were not available at the beginning. Besides, carbon dioxide loses its solubility in hot water, and so for a combination of reasons the fixing of the carbon dioxide in the primitive atmosphere could only have been a slow process.

Slow as it may have been, the fixing of carbon dioxide eventually took place. As the pressure of atmospheric carbon dioxide declined, so did the temperature at the Earth's surface. There is therefore an interesting and potentially informative relationship between early life forms and temperature, especially when we make comparisons with modern life forms, of which T. D. Brock has made valuable studies at the Yellowstone hot springs (Table 14.1). There is a strong indication that the Earth did not cool below fifty degrees centigrade until about a billion years ago, when the first eukaryotic microorganisms became evident. More complex life forms may have evolved late on the Earth for the simple reason that the temperature was previously too high.

This picture would be consistent were it not for fragmentary evidence of glaciations occurring at much more remote epochs in the pre-Cambrian era. A tillite (glacial deposit) at Gowgandra in the

Table 14.1 Approximate temperature upper limits for different groups of organisms
(after T. D. Brock)

Organism	Temperature upper limit (°C)
Nonphotosynthetic prokaryotes (bacteria)	> 90
Photosynthetic prokaryotes (blue-green algae)	73–75
Eukaryotic microorganisms (certain fungi and the alga Cyanidium caldarium)	50–60
Animals, including protozoa	45–51

Timiskaming subprovince of Ontario points to the occurrence of glaciation having occurred as long as two and a half billion years ago. Of course one might argue that the Earth has received repeated additions of volatiles, that the first fixing of carbon dioxide had indeed occurred by two and a half billion years ago, and that a subsequent high temperature phase was caused by a later acquisition of carbon dioxide. That the present picture is basically correct, however, is shown by two circumstances. In the first place, there is the correlation we have already noticed between temperature and age for the earliest life forms (Table 14.1 and Fig. 26). Secondly, there is the low

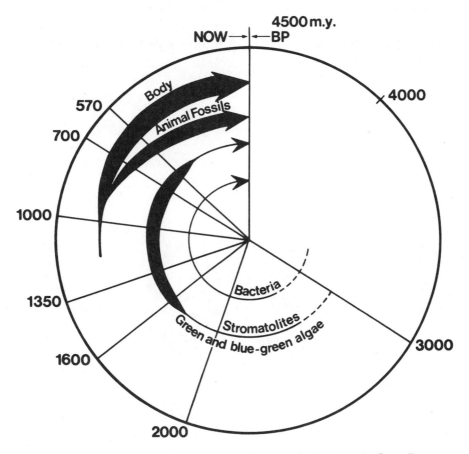

Fig. 26 Life forms in the geological record (BP = Before Present; m.y. = million years)

luminosity of the Sun. Without some strong countereffect like a high pressure of carbon dioxide, there would have been a perpetual worldwide glaciation (because of the low solar luminosity), which certainly did not happen, since the oldest rocks with ages of about 3.8 billion years are known to have been in contact with liquid water.

The best explanation therefore of the known facts relating to the origin of life on the Earth is that in the early days soft landings of comets brought about the spreading of water and other volatiles over the Earth's surface. Then about four billion years ago life also arrived from a life-bearing comet. By that time conditions on the Earth had become sufficiently similar to those on the cometary home for life to be able to persist here, probably at first tentatively and then with some assurance as time went on. The long evolution of life on the Earth had begun.

Exploring nearby planets | 15

The planets, personified or deified, pervade the mythologies of many ancient cultures. They were worshipped in the past, and more recently watched and observed with excitement and awe. The scientific exploration of the planets has been pursued with a variety of available techniques – optical, ultraviolet and infrared measurements, radar and radio astronomy – and yet some of the most startling discoveries have come from direct explorations with space craft. Our knowledge of the properties of 'near space' has changed beyond all recognition in a matter of two decades, but our knowledge about the planets, even our own planet, is far from complete, and very probably there are many surprises in store for us in the future.

The words uttered by Astronaut Neil Armstrong as he first set foot on the Moon and sank a little way into the lunar dust have become part of history: 'That's one small step for a man, one giant leap for mankind.' Whatever may be said of the cost–effectiveness of the space programme, there can be no doubt that this venture marked a most significant step in the technological progress of our species. The launching by the Russians of the first artificial satellite Sputnik 1 on 4 October 1957 set in motion a train of events that gathered momentum, mainly from a sense of competitiveness between two nations which held strongly opposed political ideologies. This led to the landing of two men on the Moon by the Americans on 20 July 1969, and to a continuing series of more ambitious efforts at exploring the solar system. The sociological impact of these events is still difficult to assess in a proper historical perspective. There may well have been some abatement of large-scale international hostilities, and an instinctive desire to colonize the globe may have become transformed into a desire for exploring the planets.

The space age has undoubtedly generated new technology, greatly improved communications, and also brought with it advances in astronomical knowledge quite far removed from probing just the solar system. There are astronomical observatories orbiting the Earth and sending us information about nebulae, stars and galaxies, in wavebands which were hitherto inaccessible from below the Earth's atmosphere. The extension of the interstellar extinction curve into the far ultraviolet (discussed in Chapter 8) was made with the use of a telescope carried on an artificial Earth satellite – and there are many other instances. In this chapter, however, our main concern will be the solar system, and in particular the question of whether there is evidence for life, or the potential for any form of life, on other planets or on planetary satellites within the solar system.

At first sight, even from ground-based studies alone, it must seem that the odds are against finding life, at any rate life that could have evolved very far, on the Moon or on any of the other planets. They all seem, at least superficially, inhospitable to life for one reason or another, which is perhaps not surprising. Two habitable planets out of nine in one solar system may be too much of a good thing, but of course we still do not know.

The Moon, to which there have been several manned flights, is the nearest and most thoroughly explored of our celestial neighbours. There have been returns of lunar samples, as well as *in situ* photographic, chemical and seismological studies, and detailed photographs have been pieced together to make up a nearly complete map of the Moon's surface including the unseen far side. Close-up pictures of the lunar landscape have been brought home by astronauts. The catalogue of discoveries revealed by direct lunar exploration is therefore an impressive record. We have the most detailed knowledge of the composition of moon rocks and moon dust as well as accurate estimates of the age of lunar surface material. The oldest rock found on the moon is 4.6 billion years old, but there appears to be very good evidence for a set of catastrophic events which re-set radioactive 'clocks' on the Moon more recently than four billion years ago. We know that there are very feeble moonquakes, and a lunar magnetic field probably imprinted by the solar wind, but there is no evidence for any atmosphere or for life. The lunar landscape is what we always suspected it to be – arid, stark and lifeless.

If, in our search for life, we move outwards from the Sun along the sequence of the planets, the first one we reach is Mercury, which is also the smallest. With a surface gravity about one third that of the Earth and with an orbital period around the Sun of eighty-eight Earth days, Mercury appears to an observer on the Earth to have phases similar to those of the Moon. From the surface of Mercury the Sun would look nearly three times larger than it does from the Earth, and the solar radiation received on a given area would be nearly ten times greater than that received by a similar area on the Earth. The period of rotation of Mercury about its axis is about fifty-nine Earth days and the surface temperature varies between about 400 degrees centigrade on the sunward side to about −200 degrees centigrade on the dark side. This intense daytime heat combined with the planet's low surface gravity is the reason why Mercury has no atmosphere beyond a tiny amount of the heavier rare gases. With a surface which probably resembles that of the Moon, Mercury is perhaps the least hospitable of the inner planets, and there can be little hope of finding life there.

Moving further out from the Sun, we come next to Venus, the planet of the goddess of love, often called the 'morning star' or 'evening star'. Venus is very similar to Earth in size and mass, and when they are at their closest the two planets are only twenty-five million miles apart. Venus has an unusual brightness which is caused by the high reflectivity of the thick clouds shrouding its surface, and it has phases like the Moon. At maximum brightness it has a brilliance which is exceeded only by the Sun and Moon. The most accurate information concerning the atmosphere of Venus came from space probes, such as the Russian space probe Venera 7, which landed on the surface on 15 December 1970 after transmitting information for twenty-three minutes and was the first unmanned spacecraft to land on another planet. Subsequent landings of Russian space probes

Venera 8, Venera 9 and Venera 10 took place in 1972 and 1976. Information from these and from earlier American Mariner probes has given us a much better idea of conditions on Venus. The composition of the atmosphere is 97 per cent carbon dioxide, 1 or 2 per cent carbon monoxide and very little oxygen or water. The top of the thick cloud cover on the planet is about forty miles above the surface. The temperature at the surface is about 500 degrees centigrade and the pressure of the atmosphere at the planet's surface is about a hundred times that of the Earth at sea level. These conditions do not look very promising for life, at any rate on the surface.

What are we to make of these surface properties of Venus? Could we rule out the existence of life higher in the atmosphere where the temperatures and pressures might be more favourable? The high surface temperature of Venus is caused by the 'greenhouse effect' due to the thick overlying layer of carbon dioxide. Sunlight filters through the atmosphere and heats the surface, but the infrared radiation emitted from the hot surface is trapped because of the greatly broadened absorption bands of carbon dioxide. The trapping of infrared radiation lifts the surface temperature in much the same way that the primeval Earth may have been heated by a carbon dioxide atmosphere, before the carbon dioxide on the Earth became fixed into limestone. The difference in the case of Venus is that there is little or no water to drive a convective refrigeration system such as we described in the last chapter. The temperature distribution in the atmosphere of Venus therefore appears to preclude the possibility of life on our neighbouring planet.

On the journey outwards from the Sun, we bypass the Earth and come to the last of the four terrestrial planets, Mars or the 'red planet' (Plates 12 and 13). For a long time it attracted more attention than any other planet because it was considered a likely habitat for intelligent life. With a radius of about half that of the Earth, and a mass of approximately one-ninth, Mars has a surface gravity which is a little less than half of terrestrial gravity. The Martian day is almost exactly as long as an Earth day, and because the tilt of its axis of rotation is the same as that of the Earth the seasons are also similar to terrestrial seasons. On the other hand, Mars is further than the Earth from the Sun, so that the Martian year is nearly twice as long as the terrestrial year.

Speculations about intelligent Martian life arose as a result of observations of many enigmatic features on the planet's surface. There are light and dark patches which show irregular as well as seasonal variations, and the dark markings were at one time widely believed to have a green colour which was attributed to vegetation. We know now, however, that such patches as appear on Mars are in fact red rather than green, and that they are caused by the redistribution of clouds of fine dust particles over the surface. There was also a long controversy over the alleged Martian 'canals'. The Italian astronomer Giovanni Schiaparelli (1835–1910) saw what he thought was an elaborate network of lines criss-crossing over the planet's surface. These so-called canals were assiduously mapped by several observers, the most notable among them being the American astronomer Percival Lowell (1855–1916).

Although evidence for intelligent Martians remained tenuous, it was at the same time difficult to disprove the theory. An argument that cannot be refuted is that if the Earth were viewed optically from a Martian vantage point, our planet would prove just as elusive over the presence or absence of intelligent life. There was no way of resolving this question unequivocally until the first Mariner probes sent back close-up pictures of the Martian surface. The answer was of course disappointingly negative. Not only was there no sign of intelligent life, but there were no structures that even resembled the fabled canals. In fact the Martian landscape looked incredibly barren and uninviting.

Mariner 9 has now surveyed the whole of the Martian surface photographically with a resolution of one kilometre, and a few per cent of the surface at a finer resolution of a hundred metres. At this latter resolution any artifacts of an intelligent civilization should have been clearly visible, but nothing of the kind was seen. Instead, it showed that the Martian surface was covered with moon-like craters and ridges. Martian craters bear a general similarity to lunar highland craters, except that they are much shallower, and they were almost certainly caused by meteoritic impacts. Some of the largest of them may be several billion years old. There is also a set of curious-looking hummocks which are a few kilometres across at the base and about a kilometre high. They are most likely to have formed by a natural physical process involving the erosion of mountains by high winds, the wind speeds on Mars being much higher than on Earth.

The Mariner probes have also shown evidence for volcanic activity at certain sites on Mars and they have given us information about the behaviour of dust storms which occur sporadically over large areas of the surface. These storms give rise to conspicuous changes in the planet's coloration as viewed from Earth. A storm which occurred in 1971 was particularly violent and widespread. In certain localized areas such as the 'dust bowl' Hellas there may be perpetual dust basins.

The equatorial surface temperature of Mars varies between a day-time high, which is close to the melting-point of water-ice, and a night-time low of about −100 degrees centigrade. (At the two landing sites of Viking 1 and Viking 2 the temperatures ranged from a high of −31 degrees centigrade to a low of −84 degrees centigrade.) The Viking 2 orbiter recorded a north pole temperature of −68 degrees centigrade. At this temperature carbon dioxide would not be frozen, so that the frozen polar cap material which was seen is believed to be water-ice.

With the possibility of an advanced life form excluded, what are the prospects for finding any form of life at all on Mars? What about humbler life forms such as microbes or algae? The physical conditions on Mars turn out to be highly restrictive, and so we cannot be over-optimistic. The Martian atmosphere is mainly comprised of carbon dioxide (about 95 per cent), argon (about 3 per cent), nitrogen (1.5 per cent) and oxygen (0.15 per cent), together with a small quantity of water. The atmospheric pressure at the surface of the planet is about half a per cent of that on Earth. Such a thin atmosphere provides no protection from the Sun's ultraviolet radiation, and the power of this flux reaching the Martian surface would prove lethal to most, if not all, terrestrial organisms. On Mars, therefore, life could exist only in niches where a natural shelter is provided from ultraviolet light. Dust basins are possible sites, since their bases may be quite well shielded by the efficient extinction properties of fine dust. It could be that organic molecules associated with pockets of microbial activity are swept up in such dust storms, and that the overlying dust protects these molecules temporarily from destruction.

A major goal of two recent probes, Viking 1 and Viking 2, which landed on Mars on 20 July and 3 September 1976, was to search for life. They were equipped to carry out biological experiments *in situ* on samples of soil, some of which were taken from under surface rocks. The presumption was that any microorganism which may be present had metabolic processes broadly similar to those of terrestrial microorganisms. The soil was treated with various nutrients, and expelled gases and the soil itself were examined in several ways. The results of the experiments turned out to be somewhat confusing. The soil was much more 'active' than any known terrestrial soil and it has been described by chemists as a super–oxidant. One curious fact is that the 'bioactivity' of the Martian soil apparently persists even after the soil is heated to well above normal sterilization temperatures. It has no detectable amounts of even simple organic compounds. This would mean that Martian microbiology, if it exists, must be remarkably thermophilic (heat-loving), and Martian ecology must provide a highly efficient scavenging system for free organic molecules. Alternatively, of course, the experiments may have been at fault. Although some of the results are marginally consistent with the existence of a Martian microbiology, scientists are careful to say that this is very far from proven. Indeed, purely chemical explanations are still thought to be preferable for most of the strange findings. The super–oxidant property of the soil is explicable, for example, in terms of the dissociation of water into free oxygen and hydrogen by solar ultraviolet light, with the light hydrogen then escaping away into space, and so leaving an excess of oxygen on Mars.

As we move further out from the Sun, prospects for life most probably recede, but we cannot be sure. Jupiter and Saturn have many 'moons' between them, and it could be that the conditions on one of these moons are favourable for life. The physical and chemical conditions on the outer planets Uranus and Neptune seem to make them especially unsuitable for life. Pluto remains an enigma about which little is known. Some astronomers think it is similar to Neptune's satellite Triton, which may well be the best 'hope' of these outermost regions of the solar system.

Finally, we can now turn from this brief survey of the life-bearing possibilities of planets and satellites to the comets themselves. Our argument suggests that it is to the comets that we must look for life, rather than to other planets or satellites of the solar system. Overwhelmingly it is in the interiors of comets where we should expect life to persist. Comets are transient, fast-moving visitors, and to dig down into one of them would be a technically difficult operation. But so great have been the advances of space research over only twenty years that perhaps we may look forward to the day when this potentially significant contribution to our understanding of the origins of life will be accomplished.

Planets of life | 16

The idea that in the whole universe life is unique to the Earth is essentially pre-Copernican. Experience has now repeatedly taught us that this type of thinking is very likely wrong. Why should our own infinitesimal niche in the universe be unique? Just as no one country has been the centre of the Earth, so the Earth is not the centre of the universe.

Before attempting to find out about other planets, we first need to know how many stars have planetary systems. In looking for an answer to this question we can be helped by our understanding of the processes which may have led to the formation of our own solar system, for the same arguments can be applied to other stars.

The starting-point of any reasonable theory of solar-system formation is a fragment of an interstellar cloud. The formation of planets is an almost inevitable consequence of the formation of a solar-type star, for although most of the cloud mass goes into the star, a small fraction eventually breaks away to form the planets. Planet formation arose from a process whereby the bulk of the star's angular momentum had to be shed along with a small fraction of its mass which went to form the planetary disc. This shedding of material had the effect of preventing a 'rotary crisis' which threatened to break up the star with too rapid a spin. This crisis does not overtake stars which are at least ten times more massive than the Sun, for these are better able to 'contain' their rotary forces. Stars of large mass are observed to spin rapidly and would not therefore be expected to have planets.

Our search for planetary systems should therefore be confined to stars less massive than about ten solar masses, though this still leaves the vast majority of stars as potential candidates. Another point about planets is that the central star, and hence the original cloud, must have basically the same chemical elements which are found in the Sun and which are essential for making up planets – oxygen, nitrogen, carbon, magnesium, silicon, iron and so on. As it happens, this condition is satisfied by the overwhelming majority of stars, so that planets with a composition like that of the Earth may well be common. Since the building-blocks of life – polysaccharides, porphyrins, nitrogen-heterocycles – are probably very widespread, potential sites for biological evolution could also be very numerous.

When it comes to the survival of life, long-term habitability and the emergence of a technological species, the conditions are much harder to fulfil. Evolution of even moderately complex life forms on Earth took over two billion years, yet a star which is fifty per cent more massive than the Sun will last only about the same time – two billion years – before swelling up to become a red-giant star. This means that there would not be enough time for life to evolve on a planet around such a star, if our terrestrial experience is regarded as anything like the norm. In looking for habitable planets we should therefore perhaps restrict our attention to stars with masses less than about one and a half times the Sun.

Stars less massive than the Sun live longer on the main sequence, which means that the red-giant phase does not intervene too early. Even so, planet-bearing stars of this type could have other restrictions and it is important to bear in mind the basic planetary requirements for habitability. In the first place, the planet must belong to a star which has a main-sequence life of not much less than three billion years, and it must have an orbit that remains substantially unperturbed over this length of time. Secondly, the average temperature on the planet must lie 'comfortably' between the melting-point of ice and the boiling-point of water. This sets a distance range from the central star which in turn depends on the star's energy output. A third condition is that the planetary mass must be within a limited range (say from about half to two and a half times an Earth mass) so as to hold an atmosphere including water – in other words an atmosphere not very different from the Earth's. In the fourth place and finally, the planet's spin on its axis should be fast enough to minimize fluctuations of the temperature between day and night. A spin period which is much longer than several Earth days would normally be too slow, for it would produce unacceptable extremes of temperature.

What about the lower mass limit of a star, if its planets are to be habitable? The distance we need to be from a furnace in order to keep us warm depends on the intensity of the heat source – the more feeble the heat source, the nearer we have to be. In a very similar way, a life-bearing planet around a star of low mass, and hence low luminosity, would have to be closer to the star in order to maintain its average surface temperature in the 'comfortable' range. For stars less massive than seventy-five per cent of a solar mass, the star-planet distance becomes small enough for the effects of tidal friction to play a dominant role. This tidal interaction has the effect of greatly slowing down the rotation of the planet on its axis, so that the resulting extremes of temperature become unacceptable for the evolution of advanced life forms. This seems to whittle down the permitted stellar mass range to between three-quarters and one and a half times that of the Sun.

We have already seen that if a planet is to be habitable, its orbit must be substantially unperturbed for at least three billion years. This means, for instance, that we could not tolerate the Earth being slowly perturbed between the orbital radii of Mercury and Jupiter. This condition would not generally be fulfilled for stars which form binary or multiple stellar systems, because these stars pursue relative motions which would tend to perturb the planetary orbits periodically. These perturbations would make conditions on planets unstable and therefore inhospitable for life.

Of the two hundred billion or so stars in our galaxy, about eighty per cent fail to meet the conditions discussed above as being necessary for life. The remaining twenty per cent are not in multiple star systems and have masses in the appropriate range, three-quarters to one and a half times the mass of the Sun. The grand total of planetary systems in the galaxy capable of supporting life is therefore close to forty billion.

With so many possible planetary systems, should we not expect inhabited planets to be moving around some of the nearby stars? We certainly should, but before expanding on this question let us take an imaginary trip to our nearest stellar neighbour, Alpha Centauri, which is about four light years away. If we were then to look back on our solar system from the vantage point of a hypothetical Earth-like planet moving around this star, the Sun would be one of the brightest stars in the sky. If Jupiter, the largest planet, were on its own it would probably be just within the margin of detectability of a 200-inch telescope such as the one on Mount Palomar in California. As it happens, from the viewpoint of Alpha Centauri, Jupiter lies only about four seconds of arc away from the Sun's point-like disc, and the solar brilliance would make detection impossible if our present-day techniques were used. In a similar way reflected light from parent stars has so far made it impossible to discover planets outside our own solar system, though such observations could be feasible in the near future with the use of large space telescopes.

Another way of detecting planets moving around a nearby star is to look for dynamical effects on the visible star produced by invisible planets. If the Sun were observed from our hypothetic vantage point near Alpha Centauri, its apparent path in the sky would show a small regular wobble with a period of about twelve years, due to a slight pull by our largest planet Jupiter. The Sun and Jupiter move around their common centre of gravity with the orbital period of 11.8 years, and this motion produces an oscillation in the Sun's position in relation to the background of more distant stars. This oscillation might conceivably be observed from Alpha Centauri.

In 1963, P. van de Kamp reported a similar effect for Barnard's star, which is the second nearest star to us. This has a mass only 0.15 times the solar mass and is 5.9 light years away from us. Van de Kamp found that the path of this star in the sky showed a wobble with a period of twenty-four years. As a result of this it was inferred that the star had an invisible planet with a mass one and a half times that of Jupiter and an orbital radius four times that of the Earth. More recently, however, studies of the path of Barnard's star have shown that there are possibly three planets. Less massive planets could also exist around Barnard's star, but these would not be detectable by the methods so far used.

There is accordingly some evidence for a planetary system around the second nearest star to us, although unfortunately the low mass and

Table 16.1 Inferred planets around Barnard's star

Planet	Mass (Jupiter = 1)	Orbital Period (year)
B$_1$	1.26	24.8
B$_2$	0.63	12.5
B$_3$	0.89	6.1

luminosity of Barnard's star probably make the planets uninhabitable according to the criteria which we have looked at. From careful studies of other nearby stars, wobbles similar to those discovered in Barnard's star have been found in a few cases, and it is possible to infer the existence of planets (Table 16.2). These inferred planets would be the most massive members of planetary families, and they can be interpreted as outermost planets which have mopped up a large fraction of hydrogen and helium from planetary discs of the type already described. In all cases, inner terrestrial planets could exist, but they remain undetected. With the exception of 70 Ophiuchi, these stars all have masses outside the range of three-quarters to one and a half solar masses. According to our criteria, 70 Ophiuchi is therefore the only candidate for a potentially 'habitable' planetary system.

A planetary system such as may exist around 70 Ophiuchi would not necessarily have a habitable planet. The probability of finding a

Table 16.2 Other nearby stars which seem to have planets (This compilation is taken from *Project Cyclops*, NASA/AMES, Report CR 11445)

Star	Distance (light years)	Mass of star (Sun = 1)	Mass of planet (Jupiter = 1)	Orbital period (years)
Cin 2347	27	0.33	20	24
70 Oph	17	0.89	10, 12	17, 10
Lac 21185	8	0.35	10	8
Kruger 60A	13	0.27	9	16
61 Cyg A	11	0.58	8	5

planet within the appropriate distance range for a 'comfortable' temperature, and with an appropriate mass range to retain a suitable atmosphere, may be about five per cent. This means that out of the forty billion possible planetary systems, only one in twenty may be congenial to life, suggesting as a very rough estimate that there may be an ultimate grand total of about two billion habitable planets in our galaxy. With a billion or so galaxies similar to our own in the observable universe, there would still be a staggering billion billion or so habitable planets in the observable universe.

The average distance between stars in our immediate vicinity is about five light years. If as few as one per cent of these are attended by habitable planets, the average distance between such planets would be about twenty-five light years. It is possible to make a list of the fourteen nearest stars to us which are most likely to possess habitable planets, and according to S. D. Dole there is chance of over forty per cent that at least one of these fourteen stars has a habitable planet (Table 16.3).

In attempting to determine the feasibility of communication with

Table 16.3 Nearest stars most likely to have habitable planets (after S. H. Dole, *Habitable Planets for Man*, Elsevier, NY 1970)

Star	Distance from Earth (light years)
Alpha Centauri A	4.3
Alpha Centauri B	4.3
Epsilon Eridani	10.8
Tau Ceti	12.2
*70 Ophiuchi A	17.3
Eta Cassiopeiae A	18.0
Sigma Draconis	18.2
36 Ophiuchi A	18.2
36 Ophiuchi B	18.2
HR 7703 A	18.6
Delta Pavonis	19.2
82 Eridani	20.9
Beta Hydri	21.3
HR 8832	21.4

(*Massive planet detected – see Table 16.2)

extraterrestrial life forms, it is necessary to estimate the average duration of a technology, though our experience in this context is of
necessity limited. In all probability our own technological development is in its infancy, for although the evolution of life on Earth has
gone on for at least three billion years, our technology – one that is
only just capable of contemplating interstellar communication – is
well under a hundred years old. If we believe the most pessimistic
prophets of doom we cannot expect our technology to last more than
three hundred years. The number of technologies at any time is given
by the formula:

Number of technologies = Number of habitable planets

$$\times \left(\frac{\text{Technological life-span}}{\text{Main-sequence life of star}} \right)$$

With two billion habitable planets and an average main-sequence life
of a star of ten billion years, this gives:

Number of technologies = Technological life-span in years ÷ 5

If three hundred years is taken as the typical technological life-span,
we obtain only sixty technologies throughout the galaxy at any time.
A much more optimistic estimate for our own technological duration
is 300,000 years, which is about the length of time homo sapiens has
been in existence. In this case the number of technologies in the galaxy
turns out to be 60,000, with a grand total of sixty million million for
the entire visible universe. Within the galaxy the average separation
between technologies would be about two hundred light years. A
direct two-way radio communication between neighbouring technologies would then take four hundred years.

There are, then, serious difficulties in calculating the number of
planets inhabited by intelligent beings in our own galaxy, let alone
throughout the observable universe. In particular, we do not know
how long our own technological civilization is going to last, or how
typical it is. Even so, we have seen that statistically there must be very
many habitable planets in our own galaxy. By looking more closely at
certain aspects of the evolution of life we should be able to find out
more about possible intelligent beings on other planets and to assess
the usefulness and methods of space travel and communication.

149

Predators and planets | 17

Wherever one travels on the Earth it will be seen that almost every bit of land where food can be grown outside urban areas has been put to use. And in the complexities of urban life every way in which a person might conceivably earn a living seems to have been found. This filling out of all possibilities is not so much a tribute to the intelligence of civilized man as it is the standard format of biological evolution, which works incessantly to seize every opportunity whereby life can be supported.

Going back to its early beginning in the solar system, life was quite likely a response to the existence of polysaccharides, chains of sugar molecules which provided an energy store to be drawn on by the first organisms. These first organisms were specialists at unzipping the chains, thereby exposing the individual molecules to 'glycosis' – the process in which sugar compounds are broken down – as for instance in a reaction that is basic to the digestion of sugars. Glucose ($C_6H_{12}O_6$) reacts with water (H_2O) to give carbon dioxide (CO_2), hydrogen (H_2) and energy:

(A) $C_6H_{12}O_6 + 6\,H_2O \rightarrow 6\,CO_2 + 12\,H_2 + \text{Energy}$

It is also likely that the bacteria in sheep and in other ruminant animals today which unzip the cellulose chains, and the enzymes which in

151

ourselves unzip the similar chains in starch, are descendants of these first organisms.

While the total quantity of polysaccharides acquired by the solar system from the interstellar gases could have been large, the amount made available to the Earth was only a small fraction of the total, and it did not last indefinitely. The eventual exhaustion of the initial energy supply from polysaccharides led to a crisis, perhaps the biggest crisis that life in the solar system has ever had to face.

The crisis was resolved when sunlight replaced polysaccharides as the basic energy source, though the use of sunlight has never been a direct one. Vision in animals, for instance, makes use of light to change the shape of a certain biomolecule, rather than to bring about a chemical reaction producing energy. In plants sunlight is used indirectly to build sugars and polysaccharides, as in the following reaction where carbon dioxide (CO_2), water (H_2O) and sunlight react to give glucose ($C_6H_{12}O_6$) and oxygen (O_2), and where glucose molecules join together to give cellulose and starch:

(B) $\quad 6\ CO_2 + 6\ H_2O + \text{Sunlight} \rightarrow C_6H_{12}O_6 + 6\ O_2$,
$\quad\quad n(C_6H_{12}O_6) \rightarrow \text{Cellulose or Starch}$

In this way sunlight, by rebuilding sugars and polysaccharides, permits life to remain substantially the way it was before.

It is here that another ingredient of plant life is involved, the biomolecule chlorophyll, for more strictly we should begin with the absorption of sunlight by the chlorophyll:

$$\text{Sunlight} + \text{Chlorophyll} \rightarrow (\text{Chlorophyll})\star$$

where the asterisk denotes energy storage by the chlorophyll. The charged–up chlorophyll drives the association of carbon dioxide and water into glucose in a reaction which we may denote by rewriting the first part of (B) in the form:

(C) $\quad 6\ CO_2 + 6\ H_2O + (\text{Chlorophyll})\star \rightarrow$
$$C_6H_{12}O_6 + 6\ O_2 + \text{Chlorophyll}$$

with the chlorophyll 'battery' becoming discharged in the formation of the glucose.

How did chlorophyll originate on the Earth? We know that chlorophyll has four heterocyclic rings (each C_4N) linked together by a magnesium atom, and with a long associated hydrocarbon side chain. We also know that the porphyrins, of which the associated heterocyclic rings of chlorophyll are an example, can absorb a great deal of light at 4430 Ångströms, a wavelength at which starlight is often found to be absorbed in its passage through the interstellar gases. This means that the basic constituents of chlorophyll may therefore well have been added to the Earth along with the initial supply of sugars. Biology therefore did not necessarily have to invent chlorophyll, which could have simply been there for the taking.

Much the same applies to retinol, the biomolecule responsible for vision by changing its shape when light hits it. It too could simply have been lying around for the taking. Vision has arisen three times during the course of biological evolution, independently for insects, the octopus and the vertebrates (including humans). In every case the basis of sight has been the changing shape of retinol. This repeated use of retinol could imply its essential superiority for vision over all other conceivable biomolecules, or it could mean that retinol was simply the most suitable of those which happened to lie to hand.

In the digestion of sugars (reaction A above) hydrogen is released, while in the formation of glucose (reaction C) oxygen is formed. Taken together, hydrogen (H_2) and oxygen (O_2) can yield water and energy in the energy-producing reaction which we ourselves experience when we breathe:

(D) $2 H_2 + O_2 \rightarrow 2 H_2O + $ Energy

The oxygen from (C) has gone into the Earth's atmosphere and we absorb it from the atmosphere into our lungs and blood, where it combines with the hydrogen we have produced internally through the glycosis of sugar, in other words through digesting our food. Once again, the energy from (D) is not used directly. Instead, its most important function is to produce adenosine triphosphate (ATP), which is mainly responsible for powering our internal metabolic processes. It is likely too that the ATP has evolved from a very early use of the element phosphorus to become the driving agent of biochemistry, and it is possible that the phosphorus itself was used from the beginning. The trend of evolution seems to have been to graft later processes onto earlier ones.

Although (A) is quite likely the earliest energy-producing

biochemical reaction and is still central to present–day life, it produces much less energy than (D). In humans, as in most multicelled animals, (A) is now mainly important as a source of hydrogen and so allows (D) to take place. The much larger energy supply from (D) enables animals to move around and it is indispensable for activities like walking, running and flying. Plants, which have (A), (B) and (C) but not (D), do not walk, run or fly, except in the imagination of science-fiction writers ignorant of chemistry.

Insects absorb oxygen directly into their bodies, leaving the oxygen to penetrate inwards by diffusion, which is a weak process that would fail to supply the central region of any really large insect, which is why insects have to be small, again except in the imagination of science-fiction writers ignorant of capillarity and gas diffusion. We and the other mammals can be big because we pump blood quickly from the lungs, where it absorbs the oxygen, to all other parts of the body, especially to the brain.

What about the culmination of evolution? Plants have a beauty and dignity of their own, but in the hierarchy of created things they are not at the top. In order to see where man stands it is worth recalling some words written by the eminent biochemist George Wald:

> We living things are a late outgrowth of the metabolism of our galaxy. The carbon and oxygen that enter so importantly into our composition were cooked in the remote past in a dying star. From them at lower temperatures nitrogen was formed. These, our indispensable elements, were spewed out into space in the exhalations of red giant [stars] and in such stellar catastrophes as supernovae, there to be mixed with hydrogen, to form eventually the substance of the Sun and planets, and ourselves. The waters of ancient seas set the pattern of ions in our blood. The ancient atmospheres molded our metabolism.
>
> We have been told so often and on such tremendous authority as to seem to put it beyond question, that the essence of things must remain forever hidden from us; that we must stand forever outside nature, like children with their noses pressed against the glass, able to look in, but unable to enter. This concept of our origins encourages another view of the matter. We are looking at it from inside. Its history is our history; its stuff, our stuff. From that realization we can take some assurance that what we see is real.
>
> Judging from our experience on this planet, such a history, that begins with elementary particles, leads perhaps inevitably toward a strange and

moving end: a creature that knows, a science-making animal, that turns back upon the process that generated him and attempts to understand it. Without his like, the universe could be, but not be known, and that is a poor thing.

Surely this is a great part of our dignity as men and women, that we can know, and that through us matter can know itself; that beginning with protons and electrons, out of the womb of time and the vastness of space, we can begin to understand; that organized as in us, the hydrogen, the carbon, the nitrogen, the oxygen, those 16 to 21 elements, the water, the sunlight – all, having become us, can begin to understand what they are, and how they came to be.

Judged from our human point of view, from the 'higher' part of us which these words of George Wald present so clearly, the process that generated us was by no means a pleasant one. To put it bluntly, it amounted to a competition to see which animal would become the greatest predator. All animals are by their nature predators, and the purpose of moving around is to widen the area over which prey may be taken. The simplest predators are the herbivores, those which feed on plants, like the ruminant animals. Plants may be said to have brought this situation on themselves, through their production of atmospheric oxygen by (C). This enabled animals, which gained an increase in energy through (D), to move around and so prey on the plants.

The aim of the predator is to cash in, biochemically speaking, on materials which some other life form has produced. The sheep eats grass in order to acquire the cellulose and starch, and hence the glucose, which it cannot produce for itself. Then the process moves to higher orders, with predators preying on other predators. As the arch-predator of the Earth, the human being preys on the sheep, which in turn preys on the grass. It all becomes a question of who can eat whom.

Our eyes are considerably more sensitive than they need to be in broad daylight, so much so that in bright conditions the eye pupil contracts in order to shut out most of the light. We are sensitive in this way because the half-hour before darkness has always been the part of the day when predators are particulary active, as when swallows wheel in their pursuit of insects during the last moments of a summer twilight. When the poet sings of the beauty of his lady's eyes he has it all wrong. Eyes did not evolve primarily because of their beauty, they evolved to prey more effectively. The modern sniper with his rifle is nearer to the natural way of things.

When we consider the full range of evolution described by George Wald, the less pleasant aspects do not apply equally everywhere. We do not think of hydrogen atoms 'suffering' in any way when they are converted into helium, and from helium into carbon, nitrogen and oxygen. In this respect, of course, the early part of evolution is quite different from the later part. The hydrogen atoms do not emit cries of agony as they cease to be, but a small bird does when it is torn to pieces by the sparrow hawk. Where does the difference lie?

The difference appears to come at precisely the place where creatures 'turn back upon the process that generated [them]' by attempting to understand it. Birds could scarcely achieve their marvels of navigation over the face of the Earth if they did not in some degree 'understand'. Science has so far failed to put a finger on where the 'turning back' process began, but quite certainly it must predate the emergence of birds. Turning back is certainly connected with the phenomenon of consciousness, but this too is imperfectly understood. Nor have we any guarantee that understanding is even possible within the limited compass of our own brains. A creature can perceive a situation without being able fully to resolve it. Birds are clearly expert at the analysis of aerial pictures, perhaps more expert than human analysts, but birds will never comprehend the mathematical theory of light or of aerodynamics. In a somewhat similar way we humans may perceive the larger pattern envisaged by George Wald, but understanding the pattern may have to await a creature more evolved than ourselves. Of course there is no certainty of this either. The full pattern may remain forever hidden, or it may be resolved by humans at some stage in our future, after a century, a millennium, or fifty thousand years. Man discovered fire a long time ago. He had to wait more than a hundred thousand years before he could understand it.

Invasion from the galaxy | 18

The point of view we have developed so far implies that the essential biochemical requirements of life exist in very large quantities within the dense interstellar clouds of gas, the so-called molecular clouds. This material became deposited within the solar system, first in comet-type bodies, and then in the collisions of such bodies with the Earth. We might speak of the Earth as having become 'infected' with life-forming materials, and other planets moving around other stars would be similarly infected. Since other planets probably exist in vast numbers – there may well be ten billion or more stars with planets in our galaxy alone – the prospect for the emergence of life on a galactic scale appears very favourable. The picture is of a vast quantity of the right kind of molecules simply looking for suitable homes, and of there being very many suitable homes.

Some scientists try to disprove this view by asking some such rhetorical question as 'Where are the alleged other planets with life?' Behind this type of question is the argument that the repeated emergence of life on many planets in the galaxy would permit further expansion of the predator-prey relationship that we discussed in the preceding chapter. The first animal to emerge anywhere in the galaxy with an adequate technology would extend its search for prey out-wards from its own stellar system to all the other systems where colonization was possible. The argument goes on to say that since the Earth must reasonably be considered suitable for colonization, and since colonization has not occurred here, as is evident from the long-continued record of biological evolution on the Earth, it follows that there can be no such other beings. Intelligent life on the Earth must therefore be essentially unique.

Although this argument contains a number of rather obviously naive assumptions, it has become widely enough discussed and believed for answers to it to be worth developing at some length. The refutation consists of two fairly separate points: the argument misunderstands the psychology of the predator-prey relationship, and colonization of the galaxy is not technically feasible.

To take the predator-prey relationship first, it is important to remember that it is not something static. There has, for instance, been a vast change in only a hundred years in the attitude of the white Australian to the aborigine. A century ago 'abos' were shot for the fun of it. Today there are vociferous conservation societies in Australia that seek to prevent the slightest interference, not only with the safety of aborigines, but with the sanctity of their territory. And today in the United States the climate of morality has changed from the attitude that 'the only good Indian is a dead Indian' to a remarkable situation in which Indians may well succeed in reclaiming much of their tribal lands throughout the State of Maine.

Such moral changes of attitude would have quite astonished people in the nineteenth century, and if such changes can happen with humans, it is perhaps not far-fetched to suppose that beings of a still higher intelligence might well have an ethic against interfering with creatures of lower intelligence such as ourselves. The cynic will smile broadly, and perhaps justifiably, at such a moral concept and will point out that never in a thousand million years of biological evolution has there been any relaxation of the severity of the predator-prey relationship. Did something quite different just happen to come along in the twentieth century? How very remarkable that would be.

Nothing new has really happened to either the white Australian-aborigine or the white American-Indian relationship. Both can be understood in terms of an age-old aspect of the attitude of predators to their prey. Predators always abandon old prey if a new and better form of prey happens to turn up. Predators optimize their search for prey quite rigorously, and this is true even for animals that are primitive compared to humans. White Australians and white Americans are now very much more moderate towards aborigines, Red Indians and blacks, mainly because there is now much less advantage in exploiting them. Conversely, white South Africans tend still to exploit their blacks, because there is still a good deal of advantage in doing so. This is an important reason for apparently ethical differences in the modern world. The cynics are right in doubting that there are ever any 'changes of heart', whatever this might mean. Changes of advantage are much more important.

The real question we have to examine therefore is what, if any, would be the advantage in colonizing the galaxy, assuming for the sake of argument that an intelligence acquired a technology adequate for the job. In particular, what would be the advantage to us, assuming we acquired a technology which enabled us to expand outwards from the Earth in a wavefront pattern (Fig. 27)? To start with, a handful of colonies would be established on the nearest suitable

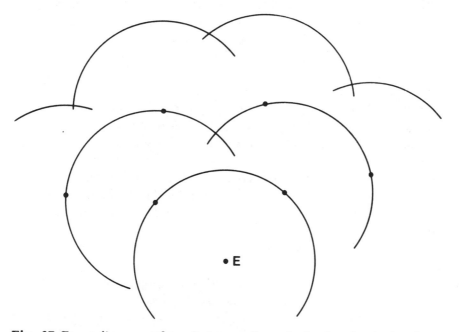

Fig. 27 Expanding wavefront in interstellar colonization, beginning from Earth (E)

planets, and from these planets further wavelets would then be sent out, followed by other wavelets, so that the ultimate effect would be a broadly expanding wavefront.

The circular arcs of Fig. 27 would probably need to have a radius of about fifty light years, which is a reasonable estimate for the spacing between planets suitable for colonization. Space-travel optimists usually calculate that would-be colonists might be able to give their vehicles a top speed of about one tenth of the speed of light, in which case the time required for each step in the expansion of the wavefront would be five hundred years. A recuperation time from one such step to another must also be considered, since the colonists would need to establish themselves before they were in a position to send out a further series of wavelets. A recuperation time of five hundred years is usually thought adequate in this respect, so that each step of the spreading wavefront of Fig. 27 would take up about a thousand years in time. The diameter of our galaxy is about a hundred thousand light years, and therefore some two thousand steps would need to be made to complete the colonization of the whole galaxy. At a thousand years per step, the time needed for the whole colonization would be about two million years, considerably longer than the human species has existed in its present form, but perhaps not so long as some superintelligence can be conceived to persist. So the argument runs.

We ourselves believe that to humans the predator-prey advantage in such a colonization process would be nil, because any prey which does not reach the predator within the predator's own lifetime is biologically useless. At no time in our history have humans deliberately sought to establish on Earth colonies that would not confer some advantage on the current generation. An immediate reward has always been expected. Indeed, it is hard to persuade humans that what is to happen only fifty years hence is of consequence to them, and a good part of the troubles of the modern world are due to our indifference in this respect.

It is not necessarily impossible, however, for intelligences to evolve with individual lifetimes appreciably longer than a thousand years, so this objection is not decisive. On the other hand, an intelligence with an individual lifetime of more than a thousand years would almost certainly have a very different view from ours of the predator-prey relationship. In view of the very different human attitudes in the nineteenth and twentieth centuries, the difference between ourselves and a postulated superintelligence would very likely be enormous. It is therefore certainly wrong to argue as if the present-day human view of the predator-prey relationship were always the norm. Furthermore, because we are of necessity uncertain what the view of a long-lived superintelligence might be, the argument becomes loose and imprecise and this is on psychological grounds alone.

The predator-prey relationship is therefore no argument against the possibility of intelligent life existing elsewhere in the universe. There are all sorts of reasons why other beings, if they exist, might not have looked upon us as prey. Even if they do exist and look upon us as prey, can they visit us? What is the technical feasibility of colonizing a galaxy? What are the possibilities and limitations of space travel by extraterrestrial beings, or indeed by us?

Space travel is made possible by advances in technology, and progress often consists of advances in the finer details of which much almost certainly remains to be discovered and invented. Even so, space technology in the future can never contradict the principles which we already know to be true, such as the principles of conservation of energy and of momentum. The thrust on a space vehicle will remain equal to the rate at which momentum is imparted to the material that exits from the vehicle. To keep the amount of such material as small as possible and so keep the overall mass to a minimum, the exit speed should be as high as possible. Exit speeds of a few kilometres per second are all that can be achieved through the combustion of chemical fuels, and the future cannot change this circumstance. Higher emission speeds can be achieved in principle in several ways, by ion-accelerators and through the emission of particles ejected in nuclear reactions. Emission speeds of the order of one tenth the speed of light, about 30,000 kilometres per second, are possibly achievable in these ways. Still higher speeds, up to that of light, are conceivable, but speeds in excess of light are impossible, another circumstance which we know to be true and which the future will not contradict.

It is essential in every moving vehicle to have means for changing the speed of the vehicle. In a car it is the accelerator or gas pedal, in an aircraft the throttles, in a space vehicle the thrust control. The rate of mass ejection from a space vehicle is connected to the thrust and to the speed u with which material is being ejected by the following simple equation:

(A)   Rate of ejection of mass = Stated thrust $\div u$

And the energy possessed by the ejected material is also connected to u and to the thrust by another very simple equation:

(B)   Rate of energy emission = Stated thrust $\times \frac{1}{2}u$

The essential difference between (A) and (B) is that the emission speed u divides in (A) and multiplies in (B). This difference creates a major problem, as we shall soon see.

The captain of a space vehicle would obviously try to conserve the amount of material available for ejection, rather as a balloonist seeks to conserve ballast. Equation (A) shows that conservation will best be achieved by adjusting the emission speed u to be as high as possible, since for a specified thrust on the vehicle the rate of mass ejection will then be least, but now (B) shows us that the rate at which energy will need to be supplied is at its maximum for the specified thrust. It is usually considered that energy supply will not in itself constitute a serious problem for the captain, because the supply can in principle be of nuclear origin. On the other hand, (B) must still present a serious problem whenever the required thrust is very large, as it must be during take-offs and landings on planetary surfaces. The trouble comes because not all the energy developed by the 'engines' of the vehicle can be carried away by the ejected material. A comparable amount of energy remains to be dissipated somehow within the vehicle itself, and if high thrust and high emission speed are associated together, the energy to be so dissipated becomes very large. Likely enough, the 'engines' cannot be cooled and the vehicle burns up.

For this reason landings and take-offs will always need to occur at low emission speed, in other words with poor efficiency. Only at low thrusts can high values of the emission speed be used. Since the interstellar parts of the journeys of would-be galactic colonizers can be at low thrust, these long parts of the journey can be at high efficiency, but the beginning and end – take-off and landing – will remain awkward. Space vehicles with capabilities for repeated landings are therefore hardly to be contemplated. Colonizers must know the planet they have chosen will be right for them before they land, since it is unlikely that there will be scope for trial and error.

Admittedly when a vehicle is reasonably close to a planet the colonizers might be expected to show good judgment of whether or not the planet would prove right for their needs, but would good judgment be possible at a distance of many light years from a planet? If the colonizers cannot make good judgments at distances of many light years, their whole project is essentially doomed to failure, as we shall now see.

Before starting on a journey, prospective colonists would almost certainly make painstaking observations of all nearby stellar systems. Equipped with a powerful space technology they would presumably make their observations with large telescopes in space, and these would presumably be far superior to our own present-day ground-based terrestrial telescopes. The colonists could reasonably be expected to determine which stellar systems possessed planets. They might well determine even the planetary sizes, masses, and distances from their respective stars, so that a short-list of 'candidate-systems' could be drawn up. But could there be absolute certainty? Because much subtler questions of biochemistry would now be involved, it may be seriously doubted whether certainty could ever be achieved.

To take an example, let us assume that a short-list of fifty candidates is drawn up with, say, an even chance that a suitable planet will be found among any ten of them. How in this case will the colonists proceed? They will set out towards the nearest candidate, which might be at a distance of, say, twenty-five light years. Once out in space, the captain accelerates at low thrust, permitting the highest attainable value of emission speed to be used – let us say that it is 30,000 kilometres per second. If the captain decides to eject half of the total mass at this speed, it can be shown that the vehicle itself attains a speed of about sixty-nine per cent of the emission speed, about 21,000 kilometres per second. Rather more than two hundred years are then needed to reach the nearest candidate.

Now the captain must have prepared his plan in the expectation that the first candidate will not prove suitable, because he will need to visit ten candidates before there is an even chance of finding a satisfactory planet. In the likelihood that the first attempt fails, what is he to do next? Let us suppose that failure becomes apparent, without a reduction in speed, as the first system is approached. If the captain tries to turn the rocket towards the further candidate that happens now to be the nearest, he will normally need to change course by a large angle. In order to do this he will need to eject a further large amount of ballast material – usually about half the remaining mass – thereby leaving only about a quarter of the original mass. As he proceeds in this way from one candidate to another, only one part in 2^n of the original mass will be left after n candidates have been visited. If ten candidates have been visited, so that $n = 10$, only one part in a thousand of the original mass will be left, and if twenty candidates ($n = 20$) should need to be visited before a suitable planet turns up, only one part in a million of the original mass will remain. Quite evidently this method of search, proceeding from one nearest candidate to another, becomes impracticable as soon as more than a very few candidates need to be visited.

There is, however, another plan open to the captain. After the first candidate has been visited, he might have enough ballast material to turn his vehicle just once through a large angle, say a hundred degrees, and enough material to slow the rocket as soon as a suitable planet is reached. The captain can then turn through, say, five degrees rather than the full hundred degrees. Since only about one part in five hundred of the 'sky' is then available, the distance of the nearest candidate in that part of the 'sky' will on the average exceed the twenty-five light years applicable for the whole sky by the cube root of five hundred. The reason for this is that the number of stars out to a given distance increases as the cube of the distance. The nearest candidate in the restricted area of the sky available to the captain can therefore be expected to be at a distance of about two hundred light years, and the time required to reach it is increased from two hundred years to about three thousand years. And since ten to twenty candidates will need to be visited, each visit requiring three thousand years, the search for a single step of the colonization process is increased to an interval of about fifty thousand years. For two thousand steps, as outlined earlier, the whole process would need some hundred million years.

While this increase still does not take the process beyond possibility, it shows how very sensitive colonization would be to restriction of the candidate list. Should it prove impracticable to restrict the list as much as we have just assumed, the search would become still more protracted and difficult. Already we need to postulate creatures with a life-span of the order of fifty thousand years, and in an implicit way we have introduced a need for space communication over big distances. From a psychological point of view, space communication would be likely to quench any desire for space travel, because space communication is so much more feasible than travel, as we shall see in our final chapter.

The need for space communication arises also because the orderly advance of the wavefront of Fig. 27 has been destroyed. After only a few steps the colonizers would become widely scattered, and without communication between them the process would become thoroughly disorganized, with suitable planets tending to be overlooked. It would, of course, be possible to return to the wavefront pattern if most space crews were prepared to accept the likelihood of complete failure. Space vehicles could be directed at every candidate planet and a few vehicles would then find colonies, but the majority would head out into space to eventual extinction. The process would be akin to the way Polynesia must have been populated, or the way seeds are scattered in the wind, with few destined to take root.

It is just here that consciousness, the 'turning-back' process of George Wald, is seen to become important. A seed has no thought or care that it may not succeed. For higher life forms, however, the urge to success and to survival is dominant, and this urge is itself a vital component of the survival process. It is true that most Polynesian pioneers must have known their chance of survival was small, but their adventures were probably not entirely of their own choosing. Many would have been committed to the ocean through being blown off course, while others would have been escaping from death at the hands of tyrants and invaders – to avoid being preyed upon. Some such motivation would be needed for a colonization of the galaxy to seem desirable. It would be hard to contemplate a fifty-thousand-year journey into the unknown with anything but a heavy heart, even for as hardy a predator as man. Since intelligence and concern for one's own well-being appear to be biologically correlated, it is even harder to imagine a superintelligence viewing such a plan with anything but repugnance.

Communicating through space | 19

Whereas space travel would demand the technology of a super-intellect, space communication lies within the compass of present-day human technology. Careful calculations, such as those made in a project studied by NASA (Project Cyclops), have shown that with radio transmitters and receivers which we ourselves could build, and with aerials no larger than the radiotelescope at Bonn, West Germany, intelligible communication could be established across a distance of about five hundred light years (Plate 14). Either by dropping the requirement (which the Bonn telescope satisfies) of being fully steer-able, or by stacking an array of many such telescopes, the communica-tion range for the resulting aerial system could be increased at least tenfold, to five thousand light years. With the technical improvements which even a very modest superintellect would surely have achieved, communication across the entire galactic system should readily be attainable.

In purely terrestrial applications, there is a substantial difference ·between a merely intelligible radio link and a high quality audio link. (A further step in quality is needed for a television link.) The difference would hardly arise for space communication, because an audio mes-sage, for example, could be transmitted more slowly than it occurs in real time. Speech could be expanded at the transmitting end and compressed back to normal speed at the receiving end. Likewise a complex, high-quality television programme could be communicated by simply transmitting at a slow enough rate. Messages would therefore need to be selective, but since thousands of years might be required for the interchange of a message there would be little demand for the constant talking which we associate with normal radio and television stations. Information in sophisticated fields, for example in mathematics, could be transmitted as rapidly as we were capable of assimilating it, and information about communications technology itself could no doubt be transmitted, partly with a view to improving quality in the sending and reception of messages.

There is a very major difference, however, between communication through a known link on the one hand and the first discovery of an initially unknown link on the other. For a known link there would be information about the point in the sky towards which we should point our aerial, and there would also be information about precisely which radio frequency we should use in the tuning of our receiver and of our transmitter. Unfortunately we ourselves have no such information. How might we go about acquiring it?

In order to find an answer to this question, let us assume that a superintellect is steadily beaming a radio message towards the solar system. Somewhere in the sky and somewhere in the radio-band a great discovery awaits' us. If, instead of steerable telescopes, fixed telescopes like the big one at Arecibo in Puerto Rico were built, each of them would be equipped with a large bank of detection horns, every one of which would be provided with radio amplification (Plate 15). The combination display from many such detection horns would be similar in principle to a photograph taken by an optical telescope. The radio 'photograph' could perhaps be made to cover a square in the sky with a side of about two degrees, so that ninety such telescopes placed at equal intervals along a line of longitude from the north geographical pole to the south pole would at any one moment cover a strip in the sky two degrees wide. As the Earth rotated, the strip would sweep across the whole sky. For one part in 180, or two in 360, of each day the message from our hypothetical superintellect would lie within one or other of ninety 'photographs' being taken by our telescopes.

The 'photographs', of course, would not be like ordinary photographs but would consist of the outputs from the radio receivers attached to the many horns sited at the focal plane of each telescope. The output from each receiver would be stored in a permanent form on magnetic tape, and so would be available at any time for computer analysis.

Next we must ask what form of 'message' we would be likely to recognize in the first place. The sequence of digits 3 1 4 1 5 9 2 could not fail to attract our immediate attention, since this is the beginning of the famous number π, the ratio of the circumference of a circle to its diameter. The digits would not necessarily be sent in the above decimal notation, however, because the superintellect would probably not know that we used the decimal system. The best notational choice for our communicator would be the binary scale, in which case the same number would take the form 1 0 1 1 1 1 1 1 1 0 1 1 1 1 1 1 0 1 1 0 0 0. The digit 1 would be represented by a radio pulse and the digit 0 by the absence of a pulse. At a reasonable transmission rate of a digit per second, only twenty-two seconds would be needed to receive this particular series of digits, whereas in a fraction of one part in 180 of a day (the time the source lies within the detection strip established by our radiotelescopes) there are 480 seconds. If constantly repeated, the number could therefore be received about twenty times every day, provided we knew the radio frequency on which to look for it – but this is the problem.

The trouble arises because the transmission must be very finely tuned, otherwise the signal from our communicator would be swamped by the general cosmic background of radio waves and by the 'noise' generated within our own terrestrial radio receivers. Contributing to the background would be the star within the communicator's own system, and this too demands that the whole power of the transmission should come out as nearly as possible to a precisely defined frequency. To minimize the background interference, the frequency chosen by our communicator would probably lie within a range from a billion cycles per second up to thirty billion cycles per second, with a variation of not more than ten cycles per second about the chosen frequency. Finding such a sharply defined frequency by random search would be rather like looking for a needle in a haystack, for a random choice of frequency would have a chance of only about one part in a thousand million of being right. If we made a new choice of frequency every twenty-two seconds, thus leaving sufficient time for each choice to recognize the starting digits of π, we could only make about twenty choices per day, so that fifty million days would have to go by before we had a good chance of landing on the right frequency. Nearly 150,000 years of search would be required.

Unless we can accurately guess the psychology of the super-intellect, or unless – and this is more likely – the superintellect has already guessed ours correctly, the needle in this particular haystack is unlikely to be found. Fortunately the relevant waveband has several precisely determined frequencies which are astronomically significant. An outstandingly important frequency occurs at 1.662 billion cycles per second, for example. This is a frequency emitted by hydroxyl radical (OH) molecules within the intersteller gas clouds. Our communicator would be manifestly unwise to attempt transmission at exactly this frequency, because there would then be competition from the molecules themselves, but a choice not far away from this frequency, or from one of the other several astronomically important frequencies, would be very sensible. The search for the right frequency would then be much restricted, perhaps as much as a thousandfold, from the unrestricted search considered in the previous paragraph. The 150,000 years of search would then be greatly shortened, to a century or two, and the problem in the simple form in which it was stated above would become soluble within a tolerable time span.

We have no knowledge, however, that our hypothetical correspondent is beaming a message towards us all the time, and indeed it would be surprising if this were so, since it is unlikely that the distant superintellect would know of our presence. Admittedly the superintellect might be sending messages in all directions, but then we can hardly suppose that the transmission for a particular direction would be as intense as it has been assumed to be in the calculations described above. There is, moreover, a difficulty in the building of an omnidirectional beacon close to a star, because the spreading of power from a particular small range of angle to all angles greatly increases the damaging radio interference from the star. Of course, we can imagine that our correspondent would take the trouble to build a radiotelescope, like that at Arecibo, for every small range of angle, and that each such telescope would be equipped with a separate transmitter. Such a program, however, would need many millions of telescopes and it hardly seems likely that the superintellect would go to such trouble to make the search problem easy for us. While it is reasonable to suppose the technology of the superintellect would exceed our own, we should not demand too much of it. If the superintellect happened to be within twenty-five light years of the Earth, it is conceivable that our own terrestrial radio transmissions might have made our presence known, and that by now a return message would already be reaching us. It is very improbable, though, that a superintellect would lie so close.

It is a sensible rule of the game not to attribute an ability to the superintellect which is manifestly outside the range that we ourselves possess. A superintellect would surely possess remarkable abilities, but it is hard, if not impossible, for us to know what they might be. Moderate extrapolations of what we ourselves can do are quite in order, but not wild guesses. It would be a moderate extrapolation to suppose that our correspondent would also build a chain of fixed radiotelescopes like the one at Arecibo, placing them at equal intervals along a line of longitude on a distant planet. Just as our telescopes would have many horns at their focal planes in order to give 'photographs' over two degree by two degree squares, so our correspondent could construct horns in similar banks at the focal plane of each telescope. Whereas, however, we would attach a radio receiver to each horn, our correspondent would attach a pulse transmitter, which is harder than attaching receivers because it involves handling a large amount of power, say a hundred kilowatts for each horn. Yet ultimately there is no great extrapolation of what we ourselves could do in supposing such a set-up, and no great measure of superintelligence would be required to achieve it.

The transmission from our correspondent at a particular moment would then cover a pole-to-pole strip with a width of two degrees. As our correspondent's planet rotated, this strip would inevitably sweep across the Earth, and it would do so for about one part in 180 – or two parts in 360 – of the time. Only during this roughly half per cent of the time would we have a chance of finding the signal. The search time needed to find the signal would therefore be increased from, say, a century up to ten thousand years. This is the length of time required to find just one communicator. Of course, if there were many communicators, the search time for finding any one of them would be significantly shortened, back again to a century if there were as many as a hundred communicators.

Nevertheless, when viewed against the immediate urgency of present-day world events, even a search time of only a century might seem an unconscionably long time, and no government or international organization is going to rush in to set up such a longitudinal chain of big radiotelescopes. Nor would it really be sensible to do so because we can well afford the time to think about the matter for a few more decades before committing ourselves to any particular system. It is generally agreed that a system based on radio, within the frequency band described above, would be better than an optical laser, but the fact that the optical laser presents itself for serious discussion at all shows that new techniques which were unthinkable before 1960 can perhaps be used to tackle the problem. It would be possible, of course, to go on for ever in such an uncertain state of mind, without a decision, on the ground that something drastically new might conceivably turn up. On the other hand, the present phase of very rapid development in communications technology, based on so-called solid-state devices, is surely exceptional. The pace of electronic development must slacken within a timescale of centuries, if not indeed within decades, and when it does so it will be time enough to draw up a firm plan for space communication.

The important thing about present-day discussion of space communication systems is to show that they can undoubtedly be achieved. The system described above, for instance, is certainly feasible. It may be cumbersome in comparison with what the future holds, but it already demonstrates the crucial point that if a potential correspondent exists anywhere within the galaxy, space communication can be established in a timescale no longer than the timescale that would in any case be necessary for an interchange of messages. The timescale of a few thousand years required for space communication is very much less than the timescale of many millions of years needed for space travel through the galaxy. Stay-at-homes will have received excellent television pictures of other creatures and of other planetary systems long before adventurous space-travellers manage to get anywhere at all. This is another answer to why other beings have never come here – space communication is so much quicker and so much easier than space travel.

Some scientists have questioned the wisdom of our becoming involved in space communication. At first sight one might think that communication could do us little harm, but such an impression may well be incorrect. Communication could possibly be as devastating as actual physical invasion, a point made by the novel *A for Andromeda*, which was a fictionalized representation of how information alone could have remarkably disruptive consequences. Here we are back at the predator-prey relationship discussed in the previous chapter, but in a new guise. Instead of being consumed physically, the prey is now to be overwhelmed intellectually. From the moment in evolution when technology provides ample access to food, the predator-prey relationship takes on a more abstract intellectual form. The game of chess is no game at all when played at a high enough level. It is a means of reducing an opponent to a grovelling wreck, which is what Bobby Fischer meant by the remark 'The weakie has made a lemon and is busted'.

The human species would be an easy pushover for an intellect even a few centuries ahead of us, let alone millions or billions of years ahead, so that the fears sometimes expressed are by no means without foundation. Yet there are salutary protections. The long timescale involved in the interchange of messages would permit recovery from intellectual defeat. New generations untainted by previous defeats would arise and would be prepared to pit their wits against an out-sider, and an intellect millions of years ahead of us would hardly take much pleasure in 'busting' us, as Bobby Fischer would hardly have taken much pleasure in 'busting' a five-year-old child. The human instinct is to be helpful whenever there is no real challenge, and the same is likely to be the case with other superior beings.

To move from the analogy of chess to English soccer, the achieve-ment of space communication would put us in the position of a club newly admitted to the fourth division. There would be an enormous way to go before we could hope to climb up among the big boys of the first division. As in soccer, we would not expect to be matched immediately against the first division, which we would hardly encounter except perhaps in an occasional cup game. The competition would be among more or less equals, as it would be all the way up the ladder.

Where else, one might ask the critic, is there for the human species to go? Nuclear bombs have essentially ruled out serious inter-human competition, even though we have not yet really grasped this fact. The land surface of the Earth is now fully explored. The Moon has been visited. A few centuries hence space travel will reach its natural limitations. What then? Degeneracy or a determined effort to climb to the first galactic division? The choice hardly needs serious discussion. Our long evolutionary experience as the Earth's outstanding predator will inevitably force us to make the attempt, and it is equally inevitable that the attempt will supply a much-needed unifying influence within our species – a unifying influence that has been so sadly lacking throughout recorded history.

Appendices

Appendix 1

THE CHEMICAL ELEMENTS AND THEIR ABUNDANCES IN THE UNIVERSE

Atomic no.	Name	Chemical symbol	Atomic weight (A)	Date of discovery	Abundance in cosmic material*
1	Hydrogen	H	1.008	1766	3.18×10^{10}
2	Helium	He	4.0003	1895	2.21×10^{9}
3	Lithium	Li	6.9	1817	49.5
4	Beryllium	Be	9.0	1798	0.81
5	Boron	B	10.8	1808	~ 10
6	Carbon	C	12.0	Old	1.18×10^{7}
7	Nitrogen	N	14.0	1772	3.64×10^{6}
8	Oxygen	O	16.0	1774	2.14×10^{7}
9	Fluorine	F	19.0	1771	2,450
10	Neon	Ne	20.2	1898	3.44×10^{6}
11	Sodium	Na	23.0	1807	6.0×10^{4}
12	Magnesium	Mg	24.3	1755	1.06×10^{6}
13	Aluminium	Al	27.0	1827	8.5×10^{5}
14	Silicon	Si	28.1	1823	10^{6}
15	Phosphorus	P	31.0	1669	9,600

*Abundances from a recent compilation by A. G. W. Cameron (*Space Science Reviews*, 15 (1970), 121–46). Notice that the abundances are *relative* to each other, with 10^6 for Si taken as the standard of reference.

Atomic no.	Name	Chemical symbol	Atomic weight (A)	Date of discovery	Abundance in cosmic material
16	Sulfur	S	32.1	Old	5.0×10^5
17	Chlorine	Cl	35.5	1774	5,700
18	Argon	A	39.9	1894	1.17×10^5
19	Potassium	K	39.1	1807	4,205
20	Calcium	Ca	40.1	1808	7.2×10^4
21	Scandium	Sc	45.0	1879	35
22	Titanium	Ti	47.9	1791	2,770
23	Vanadium	V	51.0	1830	262
24	Chromium	Cr	52.0	1797	1.27×10^4
25	Manganese	Mn	54.9	1774	9,300
26	Iron	Fe	55.9	Old	8.3×10^5
27	Cobalt	Co	58.9	1735	2,210
28	Nickel	Ni	58.7	1751	4.8×10^4
29	Copper	Cu	63.5	Old	540
30	Zinc	Zn	65.4	1746	1,245
31	Gallium	Ga	69.7	1875	48
32	Germanium	Ge	72.6	1886	115
33	Arsenic	As	74.9	Old	6.6
34	Selenium	Se	79.0	1817	67
35	Bromine	Br	79.9	1826	13.5
36	Krypton	Kr	83.3	1898	47
37	Rubidium	Rb	85.5	1861	5.88
38	Strontium	Sr	87.6	1790	26.8
39	Yttrium	Y	88.9	1794	4.8
40	Zirconium	Zr	91.2	1789	28
41	Niobium	Nb	92.9	1801	1.4
42	Molybdenum	Mo	96.0	1778	4
43	Technetium	Tc	(99)	1937	unstable
44	Ruthenium	Ru	101.1	1844	1.9
45	Rhodium	Rh	102.9	1803	0.4
46	Palladium	Pd	106.4	1803	1.3
47	Silver	Ag	107.9	Old	0.45
48	Cadmium	Cd	112.4	1817	1.42
49	Indium	In	114.8	1863	0.189
50	Tin	Sn	118.7	Old	3.59
51	Antimony	Sb	121.8	Old	0.316
52	Tellurium	Te	127.6	1782	6.41
53	Iodine	I	126.9	1811	1.09
54	Xenon	Xe	131.3	1898	5.39
55	Cesium	Cs	132.9	1860	0.387

Atomic no.	Name	Chemical symbol	Atomic weight (A)	Date of discovery	Abundance in cosmic material
56	Barium	Ba	137.4	1808	4.80
57	Lanthanum	La	138.9	1839	0.445
58	Cerium	Ce	140.1	1803	1.18
59	Praseodymium	Pr	140.9	1879	0.149
60	Neodymium	Nd	144.3	1885	0.779
61	Promethium	Pm	(147)	1947	unstable
62	Samarium	Sm	150.4	1879	0.227
63	Europium	Eu	152.0	1896	0.085
64	Gadolinium	Gd	157.3	1880	0.297
65	Terbium	Tb	158.9	1843	0.055
66	Dysprosium	Dy	162.5	1886	0.351
67	Holmium	Ho	164.9	1879	0.079
68	Erbium	Er	167.3	1843	0.225
69	Thulium	Tm	168.9	1879	0.034
70	Ytterbium	Yb	173.0	1878	0.216
71	Lutetium	Lu	175.0	1907	0.0362
72	Hafnium	Hf	178.5	1923	0.210
73	Tantalum	Ta	181.0	1802	0.0210
74	Tungsten	W	183.9	1781	0.160
75	Rhenium	Re	186.2	1925	0.0526
76	Osmium	Os	190.2	1803	0.745
77	Iridium	Ir	192.2	1803	0.717
78	Platinum	Pt	195.1	1735	1.40
79	Gold	Au	197.0	Old	0.202
80	Mercury	Hg	200.6	Old	0.40
81	Thallium	Tl	204.4	1861	0.192
82	Lead	Pb	207.2	Old	4.0
83	Bismuth	Bi	20.90	1753	0.143
84	Polonium	Po	(209)	1898	unstable
85	Astatine	At	(210)	1940	unstable
86	Radon	Rn	(222)	1900	unstable
87	Francium	Fr	(223)	1939	unstable
88	Radium	Ra	226.1	1898	unstable
89	Actinium	Ac	(227)	1899	unstable
90	Thorium	Th	232.1	1828	0.058
91	Protoactinium	Pa	(231)	1917	unstable
92	Uranium	U	238.1	1789	0.0262
93	Neptunium	Np	(237)	1940	unstable
94	Plutonium	Pu	(244)	1940	unstable
95	Americium	Am	(243)	1945	unstable

Appendix 1

Atomic no.	Name	Chemical symbol	Atomic weight (A)	Date of discovery	Abundance in cosmic material
96	Curium	Cm	(248)	1944	unstable
97	Berkelium	Bk	(247)	1950	unstable
98	Californium	Cf	(251)	· 1950	unstable
99	Einsteinium	Es	(254)	1955	unstable
100	Fermium	Fm	(253)	1955	unstable
101	Mendelevium	Md	(256)	1955	unstable
102	Nobelium	No	(253)	1958	unstable
103	Lawrencium	Lw	(256)	1961	unstable

Appendix 2

OPTICAL ISOMERISM OF BIOLOGICAL MOLECULES

Throughout this book our discussion of chemicals such as amino acids and sugars did not touch on an important property known as optical isomerism – a property which appears to be of the very essence of life. We are all of us familiar with properties of mirror images. Any structure has a mirror image which bears a likeness to the original in that component parts are similarly related within each of the structures. But in the case of an asymmetric structure the mirror image has a left-to-right transformation that cannot be reproduced by any rotation of the original structure. So the original structure and its mirror image are not identical in the same way that a right hand is not identical to a left hand.

The two amino acids L-Glycine and D-Glycine differ in the same respect – one is a mirror image of the other (Fig. 28). Similarly the two sugars, D-Glyceraldehyde and L-Glyceraldehyde are mirror images (Fig. 29).

So also do all amino acids and all sugars have distinct D- and L-forms. The physical and chemical properties of the individual D- and L-forms of a molecule are identical. They differ only in the direction in which they rotate the plane of a beam of plane polarized light (light in which vibrations of the electromagnetic field are in a single plane). Hence the term optical isomer, L referring to left-handed rotation (Levorotary) and D referring to right-handed rotation (Dextrorotary). Yet this apparently minor difference assumes a profound importance in biochemistry. Almost all amino acids

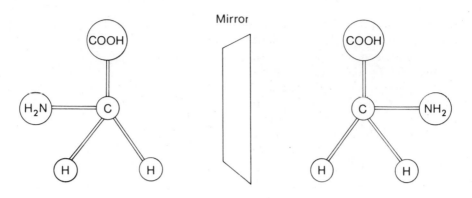

L-Glycine   **D-Glycine**

Fig. 28 Mirror images of two amino acids

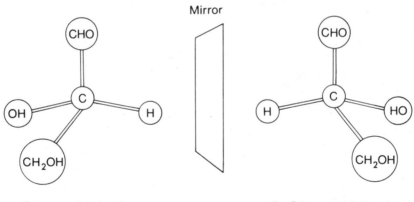

L-Glyceraldehyde   **D-Glyceraldehyde**

Fig. 29 Mirror images of two sugars

extracted from natural proteins are of the L-form and all naturally occurring polysaccharides have D-sugars. The sugar in nucleic acids is always D-Ribose. In contrast, amino acids and sugars synthesised by any non-biological procedure (e.g. the primeval soup experiments discussed in Chapter 2) are mixtures of L- and D-forms in equal proportions – what chemists call *racemic* mixture (*racemus* is the Latin for a 'bunch of grapes'). The distribution of amino acids in meteorites (discussed in Chapter 12) is also found to be *racemic*. A racemic mixture of molecules is optically inactive in the sense that it cannot rotate the plane of polarized light.

Biology has made highly specific, exclusive choices with regard to optical isomerism – D-sugars and L-amino acids. And the question is often asked – why? To begin with we note that biochemistry is often concerned not with individual molecules but with polymers or long chains of molecules. And the properties of biochemical polymers strongly depend on their shapes and on the ability of distinctively shaped molecules to fit together in biochemical reactions. We saw an illustration of this property in the action of enzymes (Chapter 5). Arbitrary replacements of L-units with D-units would completely alter the shapes and hence the properties of biochemical polymers: a left glove does not fit a right hand.

It is easy to understand why a natural polypeptide is made up of a helically wound string of amino acids all of one form. A mixture of L's and R's in the same chain makes the polymer less stable. Its growth rate would be reduced, and the length to which it can grow is also shorter. Biology would have learnt after a very few trials that to make long and stable polymers it is better to have all L's or all D's.

But why one form and not the other? Why L-amino acids, for instance? There have been many attempts to argue a case for terrestrial physical conditions producing a bias towards one direction or another, but none of these have been very convincing. It may well be that such a bias occurred already in the astronomical sources where polymers such as the polysaccharides are formed. Due to alignment of polymer chains light from the parent stars could become circularly polarised. An initial imbalance, however small, in the L's and D's in organisms which 'infected' the Earth could have led to a fight for survival as described by George Wald:

> . . . I once had the pleasure of discussing this matter with Albert Einstein. He had asked my opinion, and then said: 'You know, I used to wonder how it comes about that the electron is negative. Negative-positive – these are perfectly symmetric in physics. There is no reason whatever to prefer one to the other. Then why is the electron negative? I thought about this a long time, and at last all I could think was "It won in the fight!" ' I said, 'That's just what I think of the L-amino acids. They won in the fight!'
> In other words, if the external forces at work are indeed as symmetric as we suppose, we must assume that the original choices between optical isomers were arbitrary. At just which points such choices were made is problematical. One might well imagine a time at which there existed proteins made of either L- or D-amino acids, although for the reasons I

have given I would suppose that each such molecule tended to contain one configuration alone. Similarly, there might have been nucleic acids made with L-ribose and others made with D-ribose. It is much less likely that both groups were mixed in single organisms, even at a very primitive level, for an organism could hardly survive for any length of time if it had ceaselessly to sort out the configurations of its unit molecules in making structures of higher order, or in carrying out the chains of connected reactions upon which its metabolism depended.

Perhaps, however, there were at one time two populations of organisms, L- and D-. I am speaking loosely here of configurational relationships that hold among large series of molecules, not of directions of optical rotation, and I am also greatly oversimplifying the possibilities, which of course could include a variety of intergrades. For a time, two such enantiomorphic populations might survive together. However, as organisms began to live on other organisms, to pass material from one to another through long food chains, and to live in and on the products of other organisms, it would become highly advantageous not only for each individual organism, but for all of them collectively, to utilize single configurational series of molecules. Anything else would create endless difficulties. We see a particular instance of such difficulties in antibiotics such as gramicidin and tyrocidin that may owe their antibiotic effect, in part, to their high content of the unnatural D-configuration of amino acids.

Therefore, I should suppose that the ordinary forces of natural selection would quickly force the choice of one or the other enantiomorph, eventually, for all life on this planet. Very early in evolution certain organisms (or perhaps precursors of organisms) which, for other reasons, were superior to their enantiomorphic neighbors, won out in the struggle for existence – the 'fight' – and this decided the matter for all time . . .

. . . If the choice of optical isomers is as arbitrary as proposed, one should expect that a survey of life throughout the universe would reveal approximately equal numbers of planetary populations in which the choice of metabolically connected series of disymmetric molecules came out L- or D-; roughly equal numbers in which life is based upon L- and upon D-amino acids and, similarly, for the other molecules . . .

Index

Index

Alpha Centauri, 69
Amino acids, 47
Andromeda nebula, 61
Armstrong, Neil, 135
Autotropes, 37

Barnard's star, 146
Bell, S. J., 67
Bitz, C. M., 112
Bless, R. C., 81
Bonds, chemical, 55
Brock, T. D., 132
Brooks, J., 91
Brown, Robert, 33
Butlerov, A., 95

Carbon stars, 82
Cells
  composition of, 34
  prokaryotic and eukaryotic, 34
  structure of, 36, 152
Cellulose, 92 *et seq.*, 152 *et seq.*
Chlorophyll, 36 *et seq.*, 152
  structure of, 52, 98
Clairaut, A. C., 100

Claus, G., 110
Code, A. D., 81
Comet Kohoutek, 102
Comets, 19, 99 *et seq.*
  composition of, 103
  relation of to Uranus and
    Neptune, 101
Cosmic fireball, 60 *et seq.*
Crab nebula, 67

Danielson, R. E., 83
Darwin, Charles, 16 *et seq.*, 29 *et seq.*
DNA, 45 *et seq.*

Earth, 127 *et seq.*
  acquisition of organics and
    volatiles by, 20, 127 *et seq.*
  carbon dioxide in atmosphere of,
    131
  heating of surface of, 131
  noble gases in atmosphere of, 128
Elements
  abundances of, 42
  origin of, 57 *et seq.*
  structure of, 54

Enzymes, 38, 49 *et seq.*
Evolution, 15 *et seq.*

Fig Tree Chert, earliest living forms
    in, 31
Fischer–Tropsch synthesis, 113
Fossil record, 32
Friedmann, Alexander, 60

Galaxy, 60 *et seq.*
    colonization of, 159
    invasion from, 157
Gaustad, J. E., 83, 89
Genetic code, 48
Gillett, F. C., 89
Goldanskii, V. I., 94
Graphite particles, 82 *et seq.*

ter Haar, D., 82
Haldane, J. B. S., 25
Halley, Edmund, 100
Hartmann, J., 70
Herschel, William, 33, 70
Hertzsprung, E., 64
Hertzsprung–Russell diagram, 64
Heterotropes, 37
Hewish, A., 67
van de Hulst, H. C., 82

Iguchi, T., 113
Interstellar clouds, 70 *et seq.*
    biochemical evolution in, 77 *et seq.*
    chemical reactions in, 74
    fragmentation of, 76
    organic molecules in, 74
Interstellar dust, 79 *et seq.*, 89 *et seq.*,
    97
Isobe, S., 113

Knacke, R. F., 89

Large Magellanic Cloud, 61
Leighton, R. B., 89

Life
    on other planets, 144 *et seq.*
    origin of, 21 *et seq.*
Lindblad, B., 82
Lowell, P., 139

Magnetic fields
    in galaxy, 72
    in solar system, 121 *et seq.*
Mars, 138 *et seq.*
Meteorites, 18, 107 *et seq.*
    isotopic anomalies in, 108
    organic materials in, 112
    organized elements in, 110
    types of, 108
Microwave background radiation,
    60
Moon, 136 *et seq.*
Morimoto, M., 113
Multicelled organisms, 38

Nagakawa, N., 113
Nagy, B., 110, 112
Neugebauer, G., 89
Ney, E. P., 89
Nitrogen-bearing ring structures, 96
    et seq.
Noble gases in Earth's atmosphere,
    128
Nuclear reactions, 57 *et seq.*
Nucleic acids, 17
Nucleotide bases, 45
    pairing rules for, 46

Olavesen, A. H., 94
Oort, J. H., 105
Oparin, A. I., 25
70 Ophiuchi, 147
Orion nebula, 76

Pasteur, Louis, 23 *et seq.*
Penzias, A. A., 60

Planets
  angular momenta of, 118
  composition of, 120 *et seq.*
  condensation of, 122
  general properties of, 116
Planets in other stellar systems, 143
    et seq.
  expected frequency of, 145
  possibilities for life on, 144 *et seq.*
Polysaccharides, 43 *et seq.*, 94 *et seq.*,
    151 *et seq.*
  formation of, 95
Ponnamperuma, C., 26, 95
Porphyrins, 98   ·
Predators and their prey, 155 *et seq.*
Primordial soup, theory of, 25 *et seq.*
Project Cyclops, 167
Proteins, 47 *et seq.*
Pyran rings, 44

Radiation, 87 *et seq.*
  black-body, 87
  infrared, 88 *et seq.*
  microwave background, 60
Radio waves, synchrotron emission
    of, 72
Redi, Francesco, 23
RNA, 45 *et seq.*
Russell, H. N., 64
Rutherford, Ernest, 53

Sakata, A., 113
Schiaparelli, G., 139
Shaw, G., 91
Silicate particles, 90 *et seq.*

Solar system, 115 *et seq.*
  acquisition of organic materials
    by, 124
  angular momentum of, 118
Space communication, 167 *et seq.*
  cultural effects of, 172
  methods for achieving, 168 *et seq.*
Space exploration, 135, 161 *et seq.*
Spontaneous generation, theory of,
    23
Starlight, polarization of, 84
Stars
  lifetimes of, 65
  properties of, 61 *et seq.*
  types of, 64
Stecher, T. P., 81
Stein, W. A., 89
Sun, properties and lifetime of, 65
Supernovae, 67

Trapezium nebula, 90
Tycho Brahé, 100
Tyndall, John, 23

Universe, 59 *et seq.*
Urey, H. C., 111

Venus, 137 *et seq.*
Viruses, 39

Walcott, C. D., 31
Wald, G., 154, 165, 183
Whipple, F. L., 103
Wilson, R. W., 60
Woolf, N. J., 83